"十四五"职业教育国家规划教材

高等职业教育计算机类专业新型一体化教材

Android Studio 项目开发实战
——从基础入门到趣味开发

主　编　马　静　邝楚文　肖国金
副主编　李剑辉　李观金　刘军轶　庄焜智

电子工业出版社

Publishing House of Electronics Industry
北京·BEIJING

内 容 简 介

本书集知识性、趣味性和实用性于一体。全书以完整的 Android 项目开发流程为主线，按照项目的功能模块和 Android 项目开发的工作过程来划分工作任务，并设计教学内容，将相关知识点融入工作任务，实现理论知识与实践技能的有机结合。

本书"以项目载体为引领，以工作任务为驱动，以工作过程为基础"，共包含 9 章：Android 程序设计基础、项目简介、"登录"模块的布局、"底部导航"模块的设计、"个人中心"模块的设计、"首页"模块的设计、"吃货驾到"模块的设计、"我的订单"模块的设计、登录验证。各章内容安排合理，连贯有序，符合软件工程思想和 Android 项目开发的工作过程。读者可在完成工作任务的过程中学习 Android 项目开发的知识。

本书可作为高职高专院校 Android 项目开发课程的教学用书，也可作为各类培训机构的培训用书，以及 Android 项目开发从业人员和 Android 项目开发爱好者的参考用书。

未经许可，不得以任何方式复制或抄袭本书之部分或全部内容。
版权所有，侵权必究。

图书在版编目（CIP）数据

Android Studio 项目开发实战：从基础入门到趣味开发 / 马静，邝楚文，肖国金主编. —北京：电子工业出版社，2020.3（2024.12重印）
ISBN 978-7-121-36189-0

Ⅰ．①A… Ⅱ．①马… ②邝… ③肖… Ⅲ．①移动终端—应用程序—程序设计—高等学校—教材 Ⅳ．①TN929.53

中国版本图书馆 CIP 数据核字（2019）第 054472 号

责任编辑：李　静　　　　　　特约编辑：田学清
印　　刷：天津嘉恒印务有限公司
装　　订：天津嘉恒印务有限公司
出版发行：电子工业出版社
　　　　　北京市海淀区万寿路 173 信箱　　　邮编：100036
开　　本：787×1092　1/16　　印张：13　　字数：333 千字
版　　次：2020 年 3 月第 1 版
印　　次：2024 年12月第 17 次印刷
定　　价：43.00 元

凡所购买电子工业出版社图书有缺损问题，请向购买书店调换。若书店售缺，请与本社发行部联系，联系及邮购电话：（010）88254888，88258888。
质量投诉请发邮件至 zlts@phei.com.cn，盗版侵权举报请发邮件至 dbqq@phei.com.cn。
本书咨询联系方式：（010）88254604，lijing@phei.com.cn。

前　言

本书集知识性、趣味性和实用性于一体。全书以完整的 Android 项目开发流程为主线，按照项目的功能模块和 Android 项目开发的工作过程，将项目划分为不同的工作任务，并根据不同工作任务的特点来构建基于 Android 项目开发的工作过程的教学内容体系。本书在内容编排上按照"工作任务概述→预备知识→热身任务→任务实现"的教学思路来组织教学内容，循序渐进、由浅入深，使读者在完成工作任务的过程中学习 Android 项目开发的知识。

本书具有如下三方面的特色。

（1）"项目载体引领，工作任务驱动"。本书以培养读者的职业能力为核心，采用"项目载体引领，工作任务驱动"的教学模式，以 Android 项目开发流程为主线，按照项目的功能模块和 Android 项目开发的工作过程来划分工作任务、设计教学内容。同时以 Android 项目开发的工作过程为基础、以工作任务为驱动，实现教学过程与 Android 项目开发的工作过程对接、教学内容与岗位工作任务对接，让读者零距离体验实际的工作情景，在完成工作任务的过程中获得知识并提升技能。

（2）打破章节知识体系，重构工作过程化内容体系。本书以"基于工作过程"的职业教育思想为指导，按照实际的工作过程及人的认知心理顺序，将原本的章节知识体系打散重组，构建出基于 Android 项目开发的工作过程的教学内容体系。此外，将相关知识点融入工作任务，实现了理论知识与实践技能的有机结合。

（3）通俗易懂，趣味性强。本书用实战项目将开发中的常用技能串接起来，在各个工作任务实现的章节设置了"预备知识"和"热身任务"。"预备知识"讲解详尽，通俗易懂；"热身任务"注重趣味性和实用性，可以使读者寓学于乐、学以致用。"预备知识"和"热身任务"可以为读者对项目各工作任务的实现打好基础。同时，本书在知识讲解和任务实现过程中适时加入"小贴士"和"思考"以给予读者提示和帮助，引导读者自行思考。

本书由惠州经济职业技术学院具有丰富教学经验丰富的专业教师团队编写，凝聚了一线教师多年的课程教学经验。本书主编为马静、邝楚文、肖国金，副主编为李剑辉、李观金、刘军轶、庄煜智。感谢惠州经济职业技术学院信息工程学院薛晓萍院长及各位同事的支持和指导。

本书可作为高职高专院校 Android 项目开发课程的教学用书，也可作为各类培训机构的培训用书，以及 Android 项目开发从业人员和 Android 项目开发爱好者的参考用书。

本书配备完整的教学资源，包括电子课件、教学设计、微课、源代码等，可通过扫描以下二维码获得教学资源。

精品在线开放课程　　　电子课件、源代码、软件等　　　教学设计　　　微课下载链接

由于编者水平有限，加之编写时间仓促，书中难免存在疏漏和不足之处，恳请广大读者批评指正。编者的邮箱是 664387516@qq.com。

感谢电子工业出版社对本书的编写和出版工作给予的大力支持！

目　　录

第1章　Android 程序设计基础 1
1.1　工作任务概述 1
1.2　预备知识 1
1.2.1　基本概念 1
1.2.2　Android Studio 快速上手 2
1.2.3　Android Studio 操作界面 5
1.2.4　Android Studio 项目结构 7
1.2.5　Activity 8
1.2.6　Android 程序设计流程 8
1.2.7　Activity 的生命周期 11
1.2.8　组件的布局与属性设置 12
1.2.9　组件的事件处理 14
1.2.10　ConstraintLayout 16
1.2.11　Button 18
1.2.12　ImageView 19
1.3　搭建 Android 开发环境 19
1.4　创建并运行第一个 Android 项目 23
1.5　"滚蛋吧！肿瘤君"的界面设计 28

第2章　项目简介 32
2.1　工作任务概述 32
2.2　初识设计文档 32
2.2.1　设计文档概述 32
2.2.2　设计文档模板 32
2.3　分析开发任务 34

第 3 章 "登录"模块的布局 ... 40

3.1 工作任务概述 ... 40
3.2 预备知识 ... 41
3.2.1 View 与 ViewGroup 布局 ... 41
3.2.2 LinearLayout ... 41
3.2.3 Android 中控件的 margin 属性和 padding 属性 ... 42
3.2.4 EditText 组件 ... 42
3.2.5 Android 图片不同分辨率的适配 ... 43
3.2.6 res/values 文件夹下常用的 XML 资源文件 ... 45
3.2.7 shape ... 47
3.2.8 selector ... 52
3.3 热身任务 ... 54
3.4 实现"登录"模块的布局 ... 59

第 4 章 "底部导航"模块的设计 ... 66

4.1 工作任务概述 ... 66
4.2 预备知识 ... 67
4.2.1 Context ... 67
4.2.2 RadioGroup ... 67
4.2.3 RadioButton ... 68
4.2.4 Toast ... 69
4.3 热身任务 ... 69
4.4 实现"底部导航"模块的布局 ... 74
4.5 实现导航功能 ... 77

第 5 章 "个人中心"模块的设计 ... 82

5.1 工作任务概述 ... 82
5.2 预备知识 ... 83
5.2.1 Fragment ... 83
5.2.2 Intent ... 85
5.3 热身任务 ... 87
5.4 实现"个人中心"模块的布局 ... 92
5.5 创建"个人中心"Fragment ... 95
5.6 将"个人中心"碎片组装至 App 主框架 ... 97
5.7 实现登录界面的调用 ... 98

第 6 章 "首页"模块的设计 100

6.1 工作任务概述 100
6.2 预备知识 101
6.2.1 适配器 101
6.2.2 控件 103
6.3 热身任务 111
6.3.1 "谁是你心中的英雄" 111
6.3.2 "永不消失的经典" 113
6.3.3 "我激动,我数数" 115
6.3.4 "找不同" 118
6.4 创建"首页"Fragment 120
6.5 将"首页"碎片组装至 App 主框架 122
6.6 实现"首页"图片轮播效果 124
6.7 实现"首页"的数据适配功能 129

第 7 章 "吃货驾到"模块的设计 132

7.1 工作任务概述 132
7.2 预备知识 133
7.2.1 BaseAdapter 133
7.2.2 菜单 134
7.2.3 ContextMenu 134
7.2.4 对话框 135
7.3 热身任务 137
7.4 创建"吃货驾到"Fragment 140
7.5 将"吃货驾到"碎片组装至 App 主框架 141
7.6 实现"吃货驾到"的数据适配功能 142
7.7 实现"吃货驾到"的点赞功能 148
7.8 实现"吃货驾到"的功能菜单 149

第 8 章 "我的订单"模块的设计 152

8.1 工作任务概述 152
8.2 预备知识 153
8.3 热身任务 159
8.4 创建"我的订单"Fragment 166
8.5 将"我的订单"碎片组装至 App 主框架 168
8.6 实现"最近订单"的数据显示 169

| 8.7 | 实现"吃货驾到"的收藏功能 | 172 |
| 8.8 | 实现"我的订单"中"我的收藏"区域数据的显示 | 174 |

第9章 登录验证 178

- 9.1 工作任务概述 178
- 9.2 预备知识 179
  - 9.2.1 SharedPreferences 179
  - 9.2.2 ProgressDialog 181
  - 9.2.3 Android 网络编程 181
  - 9.2.4 用 Android 原生技术解析 JSON 185
- 9.3 热身任务 186
  - 9.3.1 "我的进度条对话框" 186
  - 9.3.2 "名人榜" 189
- 9.4 实现登录验证 194
- 9.5 实现登录信息本地保存 197

第 1 章　Android 程序设计基础

教学目标

- 了解 Android、Android Studio 及 SDK 的基本概念。
- 掌握 Android Studio 开发环境的搭建方法。
- 熟悉 Android Studio 开发环境及项目结构。
- 掌握新建及运行新项目的流程。
- 了解 Activity 的基本概念。
- 掌握 Android 程序的设计流程。
- 掌握组件布局与属性设置的方法。
- 掌握组件事件监听的添加方法。

1.1　工作任务概述

搭建 Android Studio 开发环境，创建并运行第一个 Android 项目：Hello World。在熟悉 Android Studio 开发环境的同时，学习 Android 程序的设计流程等相关知识，并完成"滚蛋吧！肿瘤君"的 UI 布局及相应功能。

1.2　预 备 知 识

1.2.1　基本概念

1. Android

Android 是一种基于 Linux 并开放源代码的操作系统，主要应用于移动设备，如智能手机和平板电脑。Android 是由 Google 和开放手机联盟领导并开发的，多称为"安卓"或"安致"。Android 最初由 Andy Rubin 开发，主要应用于手机，2005 年 8 月由 Google 收购。2007 年 11 月，Google 与 84 家硬件制造商、软件开发商及电信运营商组建开放手机联盟，共同研发并改良

Android。之后 Google 以 Apache 开源许可证的授权方式，公布了 Android 的源代码。第一部 Android 智能手机发布于 2008 年 10 月，随后，Android 的应用逐渐扩展到平板电脑及其他设备，如电视、数码相机、游戏机等。2011 年，Android 在全球的市场份额首次超过 Symbian（塞班）系统。

2. Android Studio

Android Studio 是 Google 推出的一款基于 IDEA 的 Android 集成开发工具，类似于 Eclipse ADT，它提供了集成的开发工具用于 Android 的开发和调试。

在 IDEA 的基础上，Android Studio 可以提供如下功能。

（1）基于 Gradle 的构建支持。
（2）Android 专属的重构和快速修复。
（3）提示工具，以捕获性能，解决可用性、版本兼容性等问题。
（4）支持 ProGuard 和应用签名。
（5）基于模板的向导可以生成常用的 Android 应用和组件。
（6）功能强大的布局编辑器，用户可以利用该编辑器拖拉 UI 控件并进行效果预览。

3. SDK

SDK（Software Development Kit，软件开发工具包）是为特定的软件包、软件框架、硬件平台、操作系统等建立应用软件的开发工具的集合。

4. APK

APK 是 Android Package 的缩写，即 Android 安装包。APK 是一种类似于 Symbian Sis 或 Sisx 的文件格式。APK 文件可直接安装到 Android 模拟器或 Android 智能手机中。

1.2.2 Android Studio 快速上手

Android Studio 功能强大，且其操作界面简单易用。下面对 Android Studio 进行简单介绍。

（1）一个主窗口只能打开 1 个项目，在打开第 2 个项目时，会先询问是打开当前窗口（会先关闭当前项目）还是打开新的窗口，如图 1-1 所示。若勾选图 1-1 中的 1 处则会记住用户的选择，以后不再询问；在图 1-1 中的 2 处选择是打开于当前窗口还是打开于新窗口。

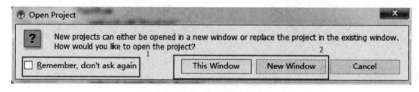

图 1-1　打开项目对话框

（2）所有对项目的变更都会自动存储，完全不需要执行存盘操作，会自动存储变更。

（3）每次启动 Android Studio 都会自动回到上一次结束时的状态。每次在打开项目时，都会出现和上一次关闭项目时相同的画面配置。若在上一次关闭 Android Studio 时没有关闭项目，则再次启动 Android Studio 时会自动打开该项目并回到上一次结束时的状态。

（4）Android Studio 有较多的快捷键，如表 1-1 所示。

表 1-1　Android Studio 常用快捷键及其功能

快　捷　键	功　　能
Ctrl+B	跳入/跳出方法或资源文件。将光标定位到某个方法或资源ID的调用处，按快捷键Ctrl+B可以跳入该方法或资源文件内部，功能等同于Ctrl+鼠标左键；如果将光标定位到方法定义处或资源文件内部，按快捷键Ctrl+B可以返回调用处
Ctrl+O	查看父类中的方法，并可以选择父类方法进行覆盖。将光标定位到类中代码的任意位置，按快捷键Ctrl+O可以在打开的面板中查看所有父类中的非私有方法，选择某个方法后按Enter键即可覆盖父类方法
Ctrl+K	SVN提交代码
Ctrl+T	SVN更新代码
Ctrl+H	查看类的上下继承关系。将光标定位到类中的任何一个位置，按快捷键Ctrl+H可以打开一个面板，在这个面板中会依照层级显示当前类的所有父类和子类
Ctrl+W	选中代码块。多次按快捷键Ctrl+W可以逐步扩大选择范围
Ctrl+E	显示最近打开的文件，可以快速再次打开这些文件
Ctrl+U	快速跳转至父类，或者快速跳转至父类中的某个方法。将光标定位到类名上，按快捷键Ctrl+U可以打开当前类的父类，如果当前类有多个父类，则会提示要打开的父类。如果一个类中的方法覆盖了其父类的方法，那么将光标定位到子类的方法，按快捷键Ctrl+U可以跳转到被覆盖的父类方法中
Ctrl+G	显示光标当前位置在代码文件中的行/列数（可以理解为光标在代码中的横纵坐标）
Ctrl+F12	查看类中的所有变量、方法、内部类、内部接口。将光标定位到当前类文件的任意位置，按快捷键Ctrl+F12可以弹出显示类中所有变量、方法、内部类、内部接口的对话框，然后按↑、↓键可以选择某个变量、方法、内部类、内部接口，接着按Enter键可以快速定位到该变量、方法、内部类、内部接口
Ctrl+F11	在光标所在行添加书签。如果文件中的代码特别多，那么书签将是一个非常实用的功能，它可以帮助用户标记代码中的重要位置，方便用户下次快速定位到这些重要位置
Shift+F11	查看书签。可以快速查看之前标记的书签
Ctrl+Shift+F12	快速调整代码编辑窗口的大小
Ctrl+↑、↓	固定光标上下移动代码
Alt+↑、↓	在内部接口、内部类和方法之间跳转
Ctrl+Shift+Backspace	回到上一次编辑的位置
Alt+数字	打开相应数字的面板。如终端控制台面板对应的数字是6，那么按快捷键Alt+6可以快速展开或关闭控制台面板
Ctrl+Shift+I	快速查看某个方法、类、接口的内容。将光标定位到某个方法、类名、接口名，按快捷键Ctrl+Shift+I可以在当前光标位置显示该方法、类、接口的内容
Shift+Esc	关闭当前打开的面板
Alt+J	选择多个相同名字的关键字、方法、类、接口，并同时对其进行更改
Ctrl+Tab	切换面板或文件。在切换面板或文件的对话框中，选中某个面板或文件，按Backspace键可以关闭该面板或文件

续表

快 捷 键	功 能
Ctrl+Shift+Enter	快速补全语句。例如，对于if(){}代码块、switch(){}代码块，只要输入if或switch（甚至sw），接着按快捷键Ctrl+Shift+Enter就可以快速完成代码块的构建
Ctrl+Alt+M	快速抽取方法。选中代码块，按快捷键Ctrl+Alt+M可以快速将选中的代码块抽取为一个方法
Ctrl+Alt+T	快速包裹代码块。选中一段代码，按快捷键Ctrl+Alt+T可以选择要对选中代码块进行的操作，如if / else、do / while、try / catch / finally等
Ctrl+Alt+L	代码格式化
Ctrl+N	快速查找类。按快捷键Ctrl+N会弹出输入类名的对话框，在该对话框的搜索框中输入要查找的类名，可以进行模糊检索，这样可以快速找到需要查找的类，这在查找类文件非常多的工程的某个或某些类时特别实用
Ctrl+Shift+N	快速查找文件。功能和Ctrl+N类似，但是除了可以搜索类文件，还可以搜索当前工程的所有文件，这是一个经常用到的、特别实用的功能
Double Shift	全局搜索。功能和Ctrl+N、Ctrl+Shift+N类似，但是搜索的范围更广，且支持符号检索，除了具有Ctrl+N、Ctrl+Shift+N的功能，还具有搜索变量、资源ID等功能
Ctrl+Alt+Space	类名或接口名提示。输入一个不完整的类名或接口名，按快捷键Ctrl+Alt+Space会给出完整类名或接口名的提示
Ctrl+Q	显示注释文档。将光标定位到某个类名、接口名或方法名，按快捷键Ctrl+Q，会显示出该类、接口、方法的注释
Ctrl+PageUp/PageDown	将光标定位到当前文件的第一行/最后一行
Shift+Left Click（当前文件的选项卡）	关闭当前文件
Ctrl+Alt+B	跳转到抽象方法的实现。将光标定位到某个抽象方法，按快捷键Ctrl+Alt+B可以快速跳转到该抽象方法的具体实现处，如果该抽象方法有多个具体实现，那么会弹出选择框进行选择
Ctrl+Shift+U	快速进行大小写转换
Ctrl+Shift+Alt+S	打开Project Structure面板
Ctrl+F	在当前文件中搜索输入的内容
Ctrl+R	在当前文件中替换输入的内容
Ctrl+Shift+F	全局搜索
Ctrl+Shift+R	全局替换
Shift+F6	快速重命名。选中某个类、变量、资源ID等，按Shift+F6可以快速重命名。只要改动一个位置该类、变量、资源ID等的名称，其他位置该类、变量、资源ID等的名称都会自动全部重命名
Alt+F7	快速查找某个类、方法、变量、资源ID被调用的地方
Ctrl+Shift+Alt+I	对项目进行审查。按快捷键Ctrl+Shift+Alt+I会弹出搜索审查项的输入框，在该输入框中输入关键字可以检索需要审查的内容。例如，输入unused resource可以搜索项目中没有使用的资源文件。此外，在菜单栏中依次单击"Analyze"→"Inspect Code"选项或右击当前工程，然后依次单击"Analyze"→"Inspect Code"选项，可以对项目进行Lint审查

续表

快 捷 键	功 能
Ctrl+D	快速复制行
Ctrl+Shift+↑、↓	上下移动代码。如果是方法中的代码，只能在方法内部移动，不能跨方法移动
Shift+Alt+↑、↓	上下移动代码。可以跨方法移动
Shift+F10	启动Module
Shift+F9	调试Module
Ctrl+F9	创建Project
Alt+Insert	快速插入代码。可以快速生成构造方法、Getter/Setter方法等
Alt+Enter	快速修复错误

1.2.3　Android Studio 操作界面

1. Android Studio 主窗口

图 1-2 为 Android Studio 主窗口，除了上方的标题、菜单栏、工具栏及底部状态栏，中间区域还有如下 3 个部分是使用频率较高的。

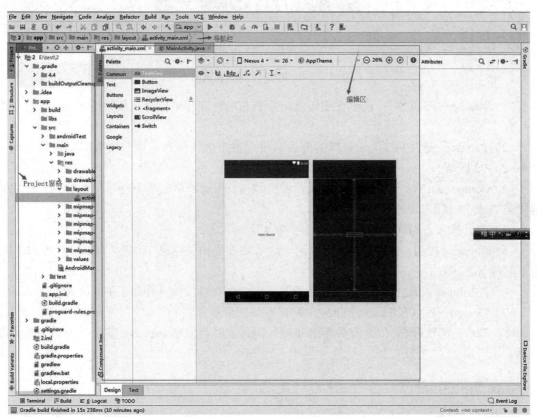

图 1-2　Android Studio 主窗口

（1）导航栏（Navigation bar）：显示当前选取或编辑中文件的路径，每一个标签表示路径中的一个文件。

（2）Project 窗格：主要用来管理项目的结构和文件。

（3）编辑区：所有打开的各类型文件都在此区域进行编辑，可利用上方的标签来切换文件。

2. Android Studio 编辑区（见图 1-3）

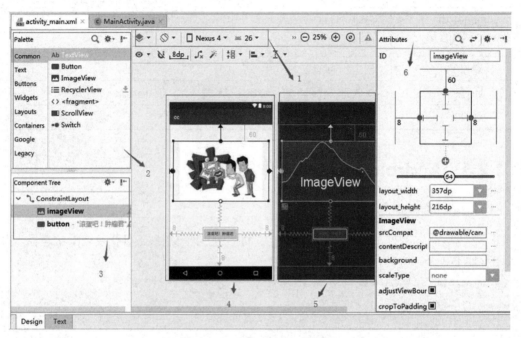

图 1-3　Android Studio 编辑区

（1）Toolbar（工具栏）：提供在编辑器中配置布局外观和编辑布局属性的按钮（图 1-3 中 1 区域）。

（2）Palette：提供小部件和布局的列表，可以将小部件和布局拖动到编辑器内的布局中（图 1-3 中 2 区域）。

（3）Component Tree：显示布局的视图层次结构。在此处单击某个项目可以看到它在编辑器中是被选中的状态（图 1-3 中 3 区域）。

（4）设计视图：用来显示界面（图 1-3 中 4 区域）。

（5）blueprint 视图：在该视图中可以清晰地看到布局相关信息、约束关系、边距等（图 1-3 中 5 区域）。

（6）Attributes（属性）：针对当前选择的视图提供属性控件（图 1-3 中 6 区域）。

（7）Design：可以在 LayoutEditor 中配置布局的外观。

（8）Text：可以查看 XML 文件中的布局代码，同时可以在 Preview 窗口中查看当前界面显示。

小贴士

在 Design 选项中，如果没有选中任何视图，在按 Ctrl 键的同时单击 UI 视图的任何位置，将会切换到 Text；如果选中了某个视图，在按 Ctrl 键的同时单击 UI 视图的任何位置，不仅会切换到 Text，而且会同时定位到选中的视图的节点在 XML 文件中的位置。

1.2.4 Android Studio 项目结构

Android Studio 项目结构图如图 1-4 所示。Android Studio 操作界面左侧的 Project 窗格用树状结构列出项目文件夹中的文件，以供用户查看和存取。由于项目内文件较多，因此 Project 窗格提供了一种 Android 查看模式，用于只显示常用文件。

图 1-4　Android Studio 项目结构图

图 1-4 中各序号对应说明如表 1-2 所示。

表 1-2　Android Studio 项目中各项资源功能说明表

图中序号	功 能 说 明
1	manifests用于存放App的配置文件
2	java用于存放程序文件和测试用的程序文件
3	res用于存放各类资源文件
4	drawable用于存放图形文件
5	layout用于存放XML布局文件
6	mipmap用于存放需要清晰显示的图形文件（如App的图标）

图中序号	功 能 说 明
7	values用于存放其他数据（如字符串、样式、尺寸等）
8	build.gradle（Project：2）中存放的是有关整个项目的Gradle配置文件
9	build.gradle（Module：app）中存放的是App模块的Gradle配置文件

1.2.5　Activity

Activity 即程序活动，简称活动，主要负责屏幕显示画面，并处理与用户的互动。Android App 是由许多画面组成的，每一个画面都由一个对应的 Activity 负责。每个 Activity 都有一个窗口画面及相对应的程序代码来处理用户与这个窗口的互动。

1.2.6　Android 程序设计流程

Android 程序的组成部分如图 1-5 所示。Android 程序设计工作大体分为两部分：一部分是程序的视觉［用户界面（User Interface），简称 UI］设计；另一部分是程序代码（程序逻辑）的编写。Android 的 UI 设计采用 XML 语言，程序代码则是用 Java 语言编写的。

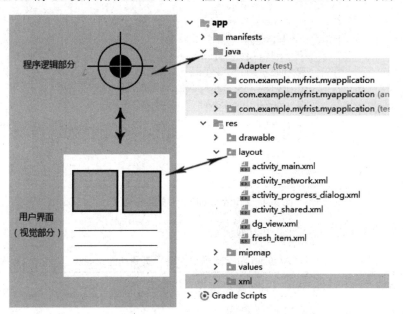

图 1-5　Android 程序的组成部分

1. 用图形化界面来进行 UI 设计

Android 采用 XML 语言来设计其 UI。Android Studio 提供了所见即所得的布局编辑器，用户只须拖动对象及设置属性即可完成 UI 布局的工作。Android Studio 会自动将用户设计好的 UI 布局转换成 XML 布局文件，该文件与 Java 程序共同构建成 App（.apk）文件。图形化界面设计如图 1-6 所示。

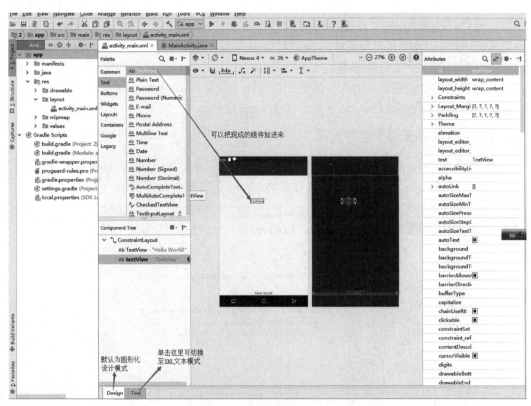

图 1-6 图形化界面设计

为了实现更好的 UI 设计效果，经常需要对 XML 布局文件进行修改。XML 布局文件示例如图 1-7 所示。

```
1.  <?xml version="1.0" encoding="utf-8"?>
2.  <android.support.constraint.ConstraintLayout
    xmlns:android="http://schemas.android.com/apk/res/android"
3.      xmlns:tools="http://schemas.android.com/tools"
4.      android:layout_width="match_parent"
5.      android:layout_height="match_parent"
6.      tools:context=".MainActivity">
7.      <TextView
8.          android:layout_width="wrap_content"
9.          android:layout_height="wrap_content"
10.         android:layout_marginBottom="292dp"
11.         android:text="Hello World!"
12.         app:layout_constraintBottom_toBottomOf="parent"
13.         app:layout_constraintHorizontal_bias="0.542"
14.         app:layout_constraintLeft_toLeftOf="parent"
15.         app:layout_constraintRight_toRightOf="parent" />
16. </android.support.constraint.ConstraintLayout>
```

图 1-7 XML 布局文件示例

图 1-7 中的相关代码功能说明如下。

第 1 行代码用来声明 XML 文件所遵循的 XML 规格版本，以及数据的编码格式。

第 2 行代码中的 xmlns:android="http://schemas.android.com/apk/res/android"主要用于设置 App 内容所需的标签。

第 3 行代码是向 Android 工具程序展示的标签。

第 4～5 行代码分别用来设置根容器的宽度与高度与其父容器的宽度与高度一致，这两行代码使得根布局的大小与设备屏幕的大小一致。

第 6 行代码用来说明当前的布局所在的渲染上下文是.MainActivity。

第 7～15 行代码表示布局中包含一个 TextView 组件。

2. 用 Java 语言来编写程序代码

Android 采用 Java 语言编写程序代码，实现相应的功能。Android Studio 为用户提供了完整的 Java 程序框架（见图 1-8），用户在建立 Android 项目时可直接引用。

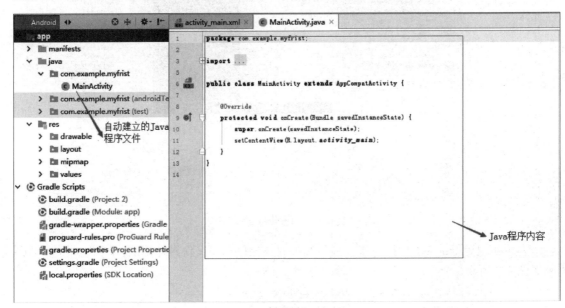

图 1-8 Java 程序框架

3. 将 UI 设计与程序代码构建（Build）成 App 文件

Android 程序设计流程图如图 1-9 所示。

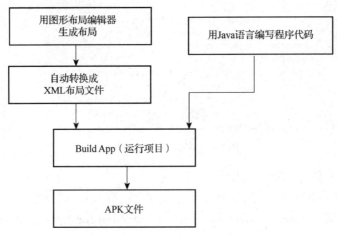

图 1-9 Android 程序设计流程图

1.2.7　Activity 的生命周期

Activity 类创建了一个窗口，开发人员可以通过 setContentView（View）接口将 UI 放到 Activity 创建的窗口中。图 1-10 为 Activity 生命周期图。

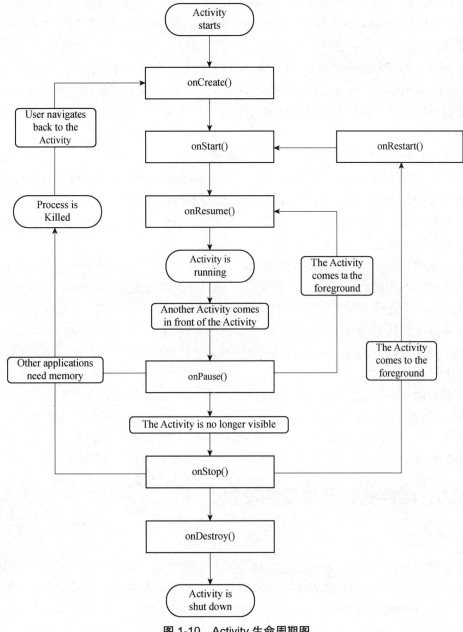

图 1-10　Activity 生命周期图

图 1-10 中的相关内容说明如下。

（1）onCreate()：当 Activity 第一次被实例化时，系统会调用该方法，且整个 Activity 生命周期只调用 1 次。onCreate()通常用于初始化设置，为 Activity 设置其所要使用的 XML 布局文件，为按钮绑定监听器等动态的设置操作。

（2）onStart()：当 Activity 可见但未获得用户焦点而不能交互时，系统会调用该方法。

（3）onRestart()：当 Activity 已经停止然后重新被启动时，系统会调用该方法。

（4）onResume()：当 Activity 可见且获得用户焦点而能交互时，系统会调用该方法。

（5）onPause()：用来存储持久数据。Activity 在这一步是可见但不可交互的，系统会停止动画等消耗 CPU 的动作。应该在这一步保存一些数据，因为这个时候程序的优先级降低，有可能被系统收回。

（6）onStop()：当 Activity 被新的 Activity 完全覆盖而不可见时，系统会调用该方法。

（7）onDestroy()：当 Activity 被系统销毁［用户调用 finish()方法或系统由于内存不足］时，系统调用该方法（整个生命周期只调用 1 次），用来释放 onCreate()方法中创建的资源，如结束线程等。

在 Activity 整个生命周期中有如下 3 个关键的循环。

（1）整个生命周期：从 onCreate()开始，到 onDestroy()结束。Activity 在 onCreate()中设置所有的"全局"状态，在 onDestory()中释放所有的资源。例如，某个 Activity 有一个在后台运行的线程，该线程用于从网络下载数据，则该 Activity 可以在 onCreate()中创建线程，在 onDestory()中停止线程。

（2）可见的生命周期：从 onStart()开始，到 onStop()结束。在可见的生命周期中，尽管 Activity 有可能不在前台，无法和用户交互但可以在屏幕上看到 Activity。在 onStart()和 onStop()之间，需要保持显示的 UI 数据和资源等。例如，可以在 onStart()中注册一个 IntentReceiver 来监听数据变化导致的 UI 变动，当不再需要显示 UI 的变动时，可以在 onStop()中注销 IntentReceiver。onStart()、onStop()都可以被多次调用，因为 Activity 随时可以在可见和隐藏之间转换。

（3）前台的生命周期：从 onResume()开始，到 onPause()结束。在前台的生命周期中的 Activity 处于所有 Activity 的最前面，与用户进行交互。Activity 可以经常性地在 Resumed 状态和 Paused 状态之间切换，如当设备准备休眠时、当一个 Activity 处理结果被分发时、当一个新的 Intent 被分发时。在这些接口方法中的代码应该属于非常轻量级别。

1.2.8　组件的布局与属性设置

1. 组件的布局

为了方便用户设计 App，Android Studio 提供了许多常用的视觉组件。用户只要将这些组件添加到 XML 布局文件的布局编辑区（或单击下方的 Text 标签切换到以文本模式加入组件的标签），就可以快速创建按钮、文本框等视觉组件。

2. 资源的 ID

当视觉和程序分开设计时，最后需要使用 R.java 文件和资源 ID 将程序与视觉组件联系起来。

在加入的新的 XML 布局文件中加入组件和图像文件等资源时，Android Studio 会自动在项目的 R.java 文件中创建代表这些资源的资源 ID，其格式为"R.资源类.资源名称"。每个资源在 R.java 文件中都有一个对应的资源 ID。因此，在程序中就可以用"R.资源类.资源名称"的格式来存取 res 文件夹下的各项资源。

双击 Shift 键，输入"R.java"就可以打开 R.java 文件，如图 1-11 所示。

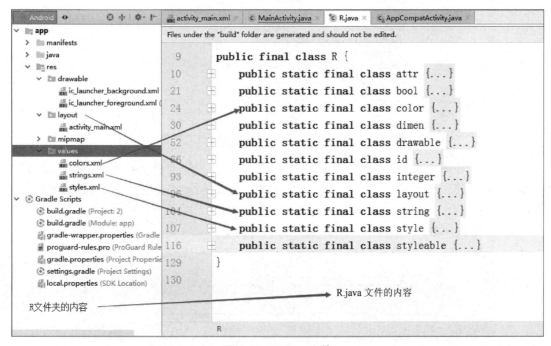

图 1-11　R.java 文件

 小贴士

1. R.java 文件中的资源都是固定的常数，其值由 Android Studio 设置，不能人工更改。

2. 在 Java 程序中是利用 findViewById()方法获取视图(布局)中的对象实体的。

3. 组件的属性设置

为了使组件达到预期的视觉效果或功能，用户需要设置组件的相关属性。下面以修改一个 TextView 组件（组件 ID 为 TextView1）的文字颜色为例来介绍常用的修改组件属性的方法。

方法一：通过修改属性控制面板中的属性值修改组件属性。选取要设置属性的 TextView 组件，在其右侧的属性控制面板中找到设置文字颜色的 textColor 属性并将其设置为#ff0000，设置完成后的效果如图 1-12 所示。

方法二：通过修改 XML 文件代码修改组件属性。将属性控制面板切换至 Text 编码方式，在 ID 为 TextView1 的组件中添加代码 android:textColor="#ff0000"，最终效果如图 1-13 所示。

图 1-12　属性控制面板

```
1  <?xml version="1.0" encoding="utf-8"?>
2  <LinearLayout xmlns:android="http://schemas.android.com/apk/res/android"
3      xmlns:app="http://schemas.android.com/apk/res-auto"
4      xmlns:tools="http://schemas.android.com/tools"
5      android:layout_width="match_parent"
6      android:layout_height="match_parent"
7      tools:context=".MainActivity">
8
9      <TextView
10         android:id="@+id/TextView1"
11         android:layout_width="wrap_content"
12         android:layout_height="wrap_content"
13         android:text="TextView"
14         android:textColor="#ff0000" />
15  </LinearLayout>
```

图 1-13 通过修改 XML 文件代码来修改文字颜色

方法三：利用 Java 代码动态修改组件属性。将属性控制面板切换至源程序编码方式，在程序中通过使用 TextView 的 tv1.setTextColor()方法动态设置文字颜色，设置完成后的效果如图 1-14 所示。

```
1. public class MainActivity extends AppCompatActivity {
2.     protected void onCreate(Bundle savedInstanceState) {
3.         super.onCreate(savedInstanceState);
4.         setContentView(R.layout.activity_main);
5.         TextView tv1=this.findViewById(R.id.textView1);
6.         tv1.setTextColor(Color.red);
7.     }
8. }
```

图 1-14 利用 Java 代码动态设置文字颜色

图 1-14 中相关代码说明如下。

第 5 行代码获取对象实体。

第 6 行代码调用 setTextColor()方法设置文字颜色。

1.2.9 组件的事件处理

当用户对手机进行各种操作时，会产生对应的事件（Event），Android 程序是通过对各种事件的处理来实现与用户的互动的。

事件发生的来源（如某个按钮）称为该事件的来源对象。如果要处理某个事件，必须准备一个能处理该事件的监听器（或称为监听对象）（Listener），当来源对象有事件发生时，就会自动调用监听对象中对应该事件的处理方法来进行处理。常用监听器如表 1-3 所示。

表 1-3 常用监听器

监 听 器	监听事件	常用组件
setOnClickListener	监听组件单击事件	Button（按钮）；ImageView（图像）；Dialog（对话框）

续表

监 听 器	监 听 事 件	常 用 组 件
setOnKeyListener	监听组件按键的各种事件（按下、弹起、保持、多次按键）	EditText（编辑框）
setOnCheckedChangeListener	监听组件选项改变事件	CheckBox（复选框）；RadioGroup（单选组）
setOnItemSelectedListener	监听组件条目获取焦点事件	Spinner（下拉列表）
setOnItemClickListener	监听组件条目单击事件	ListView（列表）；GridView（网格）
setOnDateChangedListener	监听日期选择器选择日期事件	DataPicker（日期选择器）
setOnTimeChangedListener	监听时间选择器选择时间事件	TimePicker（时间选择器）
setOnDrawerOpen（Close）Listener	监听滑动式抽屉打开（关闭）事件	SlidingDrawer（滑动式抽屉）
setOnSeekBarChangedListener	监听进度条进度变化事件	SeekBar（进度条）
setOnChronometerTickListener	监听计数器计数事件	Chronometer（计数器）
setOnTouchListenter	监听组件触屏事件（按下、弹起、保持、多次按键）	需要添加触屏功能的组件都可添加 setOnTouchListenter 监听器

为来源对象添加监听器的方法有以下 3 种（以给一个 Button 添加一个单击事件监听器为例）。

1. 方法一：匿名内部类监听单击事件

第一步：初始化控件。

```
1.   Button bt1 = (Button)findViewById(R.id.button1);
```

第二步：设置事件监听器。

```
1.   bt1.setOnClickListener(new OnClickListener(){
2.       public void onClick(View v){
3.           System.out.println("我的按钮被单击了");
4.       }
5.   });
```

2. 方法二：外部类监听单击事件

第一步：初始化控件。

```
1.   private Button bt2;   //onCreate()方法外
2.   bt2 = (Button)findViewById(R.id.button2);
```

第二步：设置事件监听器。

```
1.   bt2.setOnClickListener(new MyOnClickListener(){
2.       public void onClick(View v){
3.           super.onClick(v);   //执行父类的 onClick()方法
4.           System.out.println("我是子类");   //执行子类的 onClick()方法
5.       }
```

```
6.  });
```

父类的 onClick()方法：

```
1.  class MyOnClickListener implements OnClickListener{
2.          public void onClick(View v){
3.              System.out.println("我是父类");
4.          }
5.  }
```

 小贴士

单击按钮后会执行父类的 onClick()方法和子类的 onClick()方法，可以让多个按钮都执行相同的父类 onClick()方法。

3. 方法三：通过实现一个接口的方式实现监听单击事件

第一步：初始化组件。

```
1.  private Button bt3;  //onCreate()方法外
2.  bt3 = (Button)findViewById(R.id.button3);
```

第二步：设置事件监听器。

```
1.  bt3.setOnClickListener(this);
```

第三步：利用 MainActivity.java 类实现一个接口。

```
1.  public class MainActivity extends Activity implements OnClickListener{
2.  }
```

第四步：在 onCreate()方法外实现这个接口。

```
1.  public void onClick(View v){
2.      System.out.println("第三种方法实现");
3.  }
```

 小贴士

如果多个控件都需要设置事件监听器，那么就要使用方法三。这样只需要一个 onClick()方法，在重写的 onClick()方法中用 switch 语句来管理事件的触发，每个 case 都对应一个控件的 ID。

1.2.10　ConstraintLayout

1. ConstraintLayout 概述

ConstraintLayout（约束布局）在 Android Studio 中作为默认布局，能够减少布局的层级并改善布局性能；能够灵活地定位子 View 并调整其大小（子 View 依靠约束关系来确定位置）。一个约束关系中需要有一个 Source（源）及一个 Target（目标），Source 的位置取决于 Target。可以理解为，通过约束关系，Source 与 Target 连接在了一起，Source 相对于 Target 的位置是固定的。

2. 为视图添加一个约束

如果要为视图添加约束，首先需要将视图从 Palette 窗口拖动到布局编辑器中。当在 ConstraintLayout 中添加视图时，会在约束方框的每个角上显示一个带有方形的调整大小控件的边框，并在每条边上显示圆形约束控件，如图 1-15 所示。

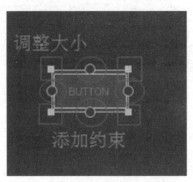

图 1-15　ConstraintLayout 调整约束示意图

图 1-16 标识了大小控件，可以拖动它来调整元素的大小。

图 1-17 标识了约束控件，单击该控件，其在元素的每一侧显示为圆形，然后将该控件拖动到另一个约束控件或父容器边界以创建约束。约束由 Z 字形线表示。

图 1-16　大小控件　　　　　　　　图 1-17　约束控件

注意：在为控件添加约束条件时，最少需要添加两个约束条件（左上、左下、右上、右下），当添加的约束条件不满足时，会出现报红线的控件。

3. ConstraintLaytout 常用属性及说明（见表 1–4）

表 1-4　ConstraintLaytout 常用属性及说明

属　　性	说　　明
app:layout_constraintTop_toTopOf	将所需视图的顶部与另一个视图的顶部对齐
app:layout_constraintTop_toBottomOf	将所需视图的顶部与另一个视图的底部对齐
app:layout_constraintBottom_toTopOf	将所需视图的底部与另一个视图的顶部对齐
app:layout_constraintBottom_toBottomOf	将所需视图的底部与另一个视图的底部对齐
app:layout_constraintLeft_toTopOf	将所需视图的左侧与另一个视图的顶部对齐
app:layout_constraintLeft_toBottomOf	将所需视图的左侧与另一个视图的底部对齐
app:layout_constraintLeft_toLeftOf	将所需视图的左侧与另一个视图的左侧对齐
app:layout_constraintLeft_toRightOf	将所需视图的左侧与另一个视图的右侧对齐

属　　性	说　　明
app:layout_constraintRight_toTopOf	将所需视图的右侧与另一个视图的顶部对齐
app:layout_constraintRight_toBottomOf	将所需视图的右侧与另一个视图的底部对齐
app:layout_constraintRight_toLeftOf	将所需视图的右侧与另一个视图的左侧对齐
app:layout_constraintRight_toRightOf	将所需视图的右侧与另一个视图的右侧对齐

ConstraintLayout 的通用形式如下。

app:layout_constraintXXX_toYYYOf ="@+id/view"：表示将所需 A 视图的 X 方位与 B 视图的 Y 方位对齐。

当 XXX 和 YYY 相反时，表示组件自身的 XXX 在约束组件的 YYY 一侧。例如，app:layout_constraintLeft_toRightOf="@id/button1"表示的是组件自身的左侧在 button1 的右侧。

当 XXX 和 YYY 相同时，表示组件自身的 XXX 和约束组件的 YYY 的一侧对齐。例如，app:layout_constraintBottom_toBottomOf="parent"表示组件自身底部与父组件底部对齐。

ConstraintLayout 方位示意图如图 1-18 所示。

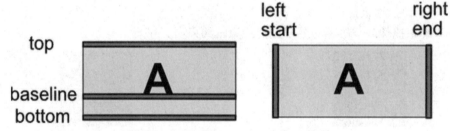

图 1-18　ConstraintLayout 方位示意图

1.2.11　Button

Button（按钮）是各种 UI 中常用的组件之一，用户可以通过触摸它来触发一系列事件，如单击事件等。Button 的常用属性及说明如表 1-5 所示。

表 1-5　Button 的常用属性及说明

属　　性	说　　明
android:clickable	设置是否允许单击。 clickable=true：允许单击 clickable=false：禁止单击
android:background	通过资源文件设置背景色。 按钮默认背景色为android.R.drawable.btn_default
android:text	设置文字
android:textColor	设置文字颜色
android:onClick	设置单击事件

1.2.12 ImageView

ImageView（图片）组件负责显示图片，其图片的来源既可以是资源文件的 ID，也可以是 Drawable 对象或 Bitmap 对象，还可以是 Content Provider 的 URL。ImageView 的常用属性及说明如表 1-6 所示。

表 1-6　ImageView 的常用属性及说明

属　　性	说　　明
android:layout_width	组件宽
android:layout_height	组件高
android:scaleType	组件如何显示。 相关参数说明如下。 center：按图片原来的尺寸居中显示，当图片的长（宽）超过View的长（宽）时，则截取图片居中部分显示； centerCrop：按比例扩大图片的尺寸后居中显示，使图片的长（宽）等于或大于View的长（宽）； centerInside：将图片的内容完整居中显示，按比例缩小图片尺寸后或按原来的尺寸使图片的长（宽）小于或等于View的长（宽）； fitCenter：将图片按比例扩大/缩小到View的宽度，居中显示； fitEnd：将图片按比例扩大/缩小到View的宽度，显示在View的下半部分； fitStart：将图片按比例扩大/缩小到View的宽度，显示在View的上半部分； fitXY：将图片按比例扩大/缩小到View的大小并显示； matrix：用矩阵来绘制
android:src	设置显示图片

1.3　搭建 Android 开发环境

1. 知识点

- JDK 的下载、安装及配置方法。
- Android Studio 的下载、安装及配置方法。

2. 工作任务

搭建基于 Android Studio 的 Android 开发环境。

3. 操作流程

（1）下载及安装 JDK。要下载 Oracle 公司的 JDK，JDK 下载页面如图 1-19 所示，在此页面中选择与自己的计算机系统对应的版本即可。

将 JDK 下载到本地计算机后对其进行安装。JDK 在默认安装地址中安装成功后，会在系统目录下出现两个文件夹，一个代表 jdk，另一个代表 jre。JDK 的全称是 Java SE Development Kit，即 Java 开发工具箱，SE 表示标准版。JDK 是 Java 的核心，包含 Java 运行环境（Java Runtime

Environment)、Java 开发工具和开发者开发应用程序时所要调用的 Java 类库。

图 1-19　JDK 下载页面

（2）配置 JDK 的变量环境。JDK 环境变量配置窗口如图 1-20 所示。利用 Windows 环境变量设置工具，分别添加 JAVA_HOME、PATH 和 CLASSPATH 这 3 个变量，具体设置方法如下。

图 1-20　JDK 环境变量配置窗口

①JAVA_HOME：变量值为 JDK，安装路径为 C:\Program Files\Java\jdk1.9.1。该变量创建完成后可以利用%JAVA_HOME%作为 JDK 安装目录的统一引用路径。

②PATH：由于 PATH 属性本就存在，因此可直接对其进行编辑，在原来的变量后添

加%JAVA_HOME%\ bin;%JAVA_HOME%\jre\bin 即可。

③CLASSPATH：设置系统变量名为 CLASSPATH，变量值为.;%JAVA_HOME%\lib\dt.jar;%JAVA_HOME%\lib\tools.jar。注意变量值字符串前面有一个"."，表示当前目录。设置 CLASSPATH 的目的在于告诉 Java 执行环境，在哪些目录下可以找到所要执行的 Java 程序所需要的类或包。

（3）测试 JDK 是否安装成功。依次单击"开始"→"运行"按钮，输入 cmd，打开命令行模式。输入命令 java-version 检测 JDK 安装是否成功。如果运行结果如图 1-21 所示，则表示 JDK 安装成功。

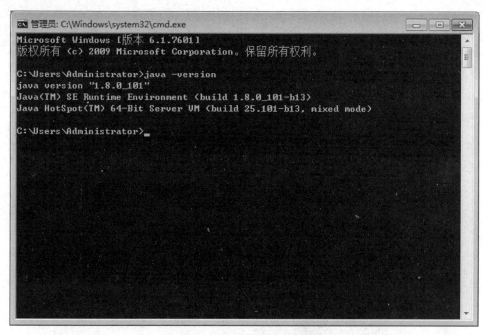

图 1-21 JDK 安装成功示意图

（4）下载 Android Studio。在 Android Studio 中文社区选取相应的 Android Studio 安装包进行下载。

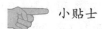 小贴士

在 Android Studio 中文社区下载页面，有些 Android Studio 安装包不包含 Android SDK。在这种情况下，用户需要另外下载 Android SDK，这样很容易出现各种兼容性问题。故强烈建议初学者要下载和使用集成有 Android SDK 的 Android Studio。

（5）执行 Android Studio 的安装向导。Android Studio 下载完成后，启动安装向导。安装向导开始之后，一直单击"Next"按钮，直到出现选择组件窗口。勾选全部组件复选框，如图 1-22 所示。然后单击"Next"按钮。之后再次同意条款和条件，当出现如图 1-23 所示窗口时，可以选择 Android Studio 的安装位置。

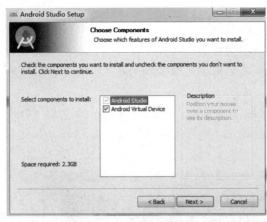
图 1-22　选择组件窗口

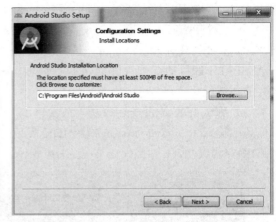
图 1-23　选择 Android Studio 的安装位置

选择好 Android Studio 的安装位置后，单击 "Next" 按钮，直到出现如图 1-24 所示的窗口。勾选 "Start Android Studio" 复选框并单点击 "Finish" 按钮，启动 Android Studio。

图 1-24　Android Studio 安装完成

在第一次打开 Android Studio 时，安装向导将查找该计算机系统中的 JDK 和 Android SDK 的位置。安装向导会为 Android Studio 下载开发应用程序需要的东西。单击 "Finish" 按钮关闭安装向导。

图 1-25　安装向导（下载组件）

1.4 创建并运行第一个 Android 项目

1. 知识点
- ➢ Android Studio 的启动过程、集成环境的基本组成、菜单组成与工具栏。
- ➢ 使用 Android Studio 创建新项目的方法。
- ➢ 使用 Android Studio 运行项目的方法。

2. 工作任务

在基于 Android Studio 的 Android 开发环境中创建项目 Hello World，并分别在真机和模拟器上运行。

3. 操作流程

（1）当安装向导运行完成后，将会出现 Android Studio 的欢迎界面，如图 1-26 所示，在该界面中单击 "Start a new Android Studio project" 选项。

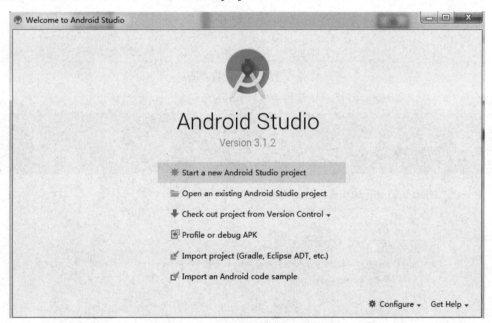

图 1-26　Android Studio 欢迎界面

（2）在新建项目界面中的 "Application name" 中输入 Hello World，在 "Company domain" 中输入 myfrist.example.com，如图 1-27 所示。需要注意的是，"Package name" 是反转的 Company Domain 加上 Application name。然后在 "Project location" 中设置项目保存路径。

（3）Android 可以运行于多种平台，包括游戏机、电视机、手表、眼镜、智能手机和平板电脑。在如图 1-28 所示界面勾选所有复选框，然后单击 "Next" 按钮。

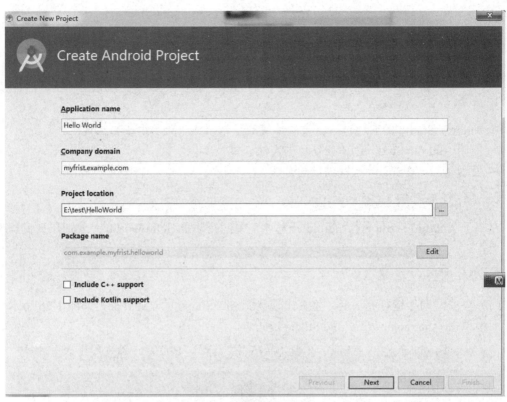

图 1-27 新建项目界面

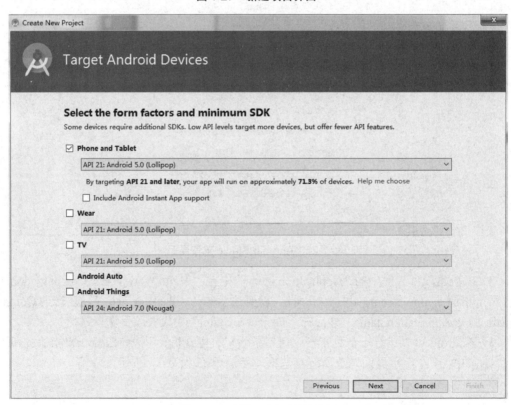

图 1-28 选择项目运行的形式

（4）新建项目向导将提示用户选择一种 Activity，这里选择 Empty Activity 并同意默认名字，然后单击"Next"按钮，如图 1-29 所示。

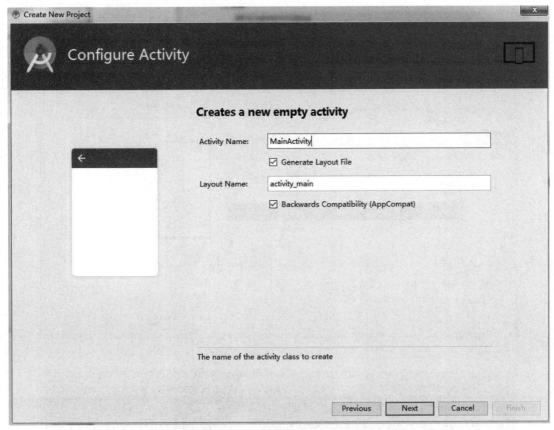

图 1-29 为新建项目选择 Activity

（5）一直单击"Next"按钮，当出现"Finish"按钮时单击该按钮便可完成一个新项目的创建。

（6）创建模拟器。在 Android 虚拟设备管理器向导的第一个界面中单击 AVD（创建 Android 虚拟设备）图标，如图 1-30 所示。在出现的下一个界面选择"Galaxy Nexus"选项，如图 1-31 所示，然后单击"Next"按钮。在出现的下一个界面允许选择一个系统镜像，如图 1-32 所示，在该界面选择第一个选项"Lollipop"（或最新的 API）和"x86_64"系统镜像并单击"Next"按钮。在出现的下一个界面中，单击"Finish"按钮来验证 AVD 设置。此时，已经创建了一个新的 AVD。

图 1-30 AVD 图标

 小贴士

如果需要创建一个 Android Studio 还没有定义的设备，建议通过 phonearena.com 查找具

体型号。之后，使用 1.4 节介绍的相关步骤创建一个新的 AVD。Genymotion 是一款优秀的、第三方市场的 Android 模拟器，它对非商业目的用户是免费的，而且使用非常方便。

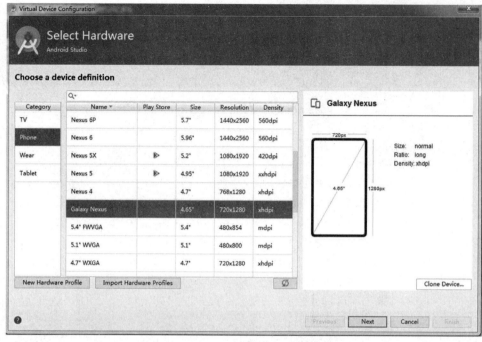

图 1-31　选择 Galaxy Nexus

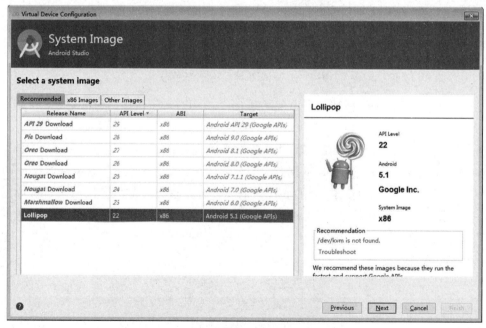

图 1-32　选择 Lollipop

（7）在 AVD 上运行 Hello World 应用程序。在新创建的 AVD 上运行 Hello World 应用程序，单击工具栏中的"运行"按钮，如图 1-33 所示。

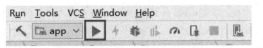

图 1-33 "运行"按钮

确保"Launch emulator"单选按钮被选中,并选择"Galaxy Nexus API 21"选项,然后单击"OK"按钮,如图 1-34 所示。当 AVD 启动后,能够看到 Hello World 应用程序运行在模拟器上,如图 1-35 所示。

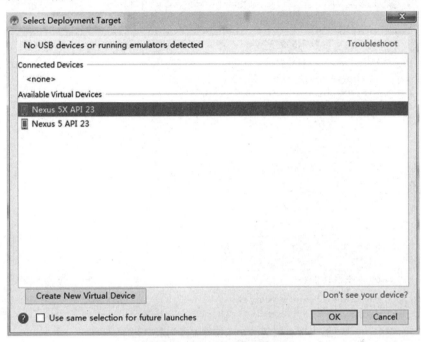

图 1-34 选择设备开启模拟器

图 1-35 Hello World 应用程序运行在模拟器上

（9）在 Android 设备上运行 Hello World 应用程序。AVD 对于模拟特定的设备是非常实用的，尤其当没有相关的物理设备时。当条件允许时，首选在物理设备上开发应用程序。Android 设备通过数据线与计算机相连，单击"运行"按钮（见图 1-33），如果驱动程序安装正确，能够看到列出来的已连接的物理设备，如图 1-36 所示。在图 1-36 所示界面选择已连接的 Android 设备，单击"OK"按钮，等待几秒钟后可以看到 Hello World 应用程序运行在 Android 设备上了。

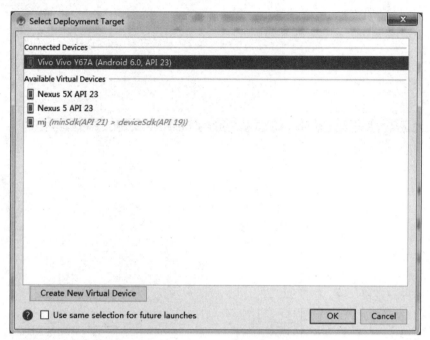

图 1-36　在 Android 物理设备列表中选择设备

 1.5　"滚蛋吧！肿瘤君"的界面设计

1. 知识点

➢ Android 程序的设计流程。
➢ Button 组件。
➢ ImageView 组件。
➢ 单击监听的添加。
➢ 修改属性的方法。

2. 工作任务

（1）完成如图 1-37 所示的界面布局。
（2）当单击图 1-37 中的"滚蛋吧！肿瘤君"按钮后，界面中的图片被隐藏，效果如图 1-38 所示。

图 1-37　单击"滚蛋吧！肿瘤君"按钮前　　图 1-38　单击"滚蛋吧！肿瘤君"按钮后

3. 操作流程

（1）打开 Android Studio，新建一个名为 cancer 的项目。
（2）将图片素材 cancer.jpg 复制到项目中的 res/drawable 文件夹中。
（3）打开项目中的 res/layout 文件夹下的 activity_main.xml 布局文件，并在该布局文件中添加 1 个 ImageView 组件、1 个 Button 组件，布局效果如图 1-39 所示。

图 1-39　"滚蛋吧！肿瘤君"布局效果

（4）修改各组件属性。修改后的 activity_main.xml 布局文件代码如下。

```
1.  <?xml version="1.0" encoding="utf-8"?>
2.  <android.support.constraint.ConstraintLayout
        xmlns:android="http://schemas.android.com/apk/res/android"
3.      xmlns:app="http://schemas.android.com/apk/res-auto"
4.      xmlns:tools="http://schemas.android.com/tools"
5.      android:layout_width="match_parent"
6.      android:layout_height="match_parent"
7.      tools:context=".MainActivity">
8.      <ImageView
9.          android:id="@+id/imageView"
10.         android:layout_width="357dp"
11.         android:layout_height="216dp"
12.         android:layout_marginEnd="8dp"
13.         android:layout_marginStart="8dp"
14.         android:layout_marginTop="60dp"
15.         app:layout_constraintEnd_toEndOf="parent"
16.         app:layout_constraintHorizontal_bias="0.545"
17.         app:layout_constraintStart_toStartOf="parent"
18.         app:layout_constraintTop_toTopOf="parent"
19.         app:srcCompat="@drawable/cancer" />
20.     <Button
21.         android:id="@+id/button"
22.         android:layout_width="131dp"
23.         android:layout_height="wrap_content"
24.         android:layout_marginBottom="8dp"
25.         android:layout_marginEnd="8dp"
26.         android:layout_marginStart="8dp"
27.         android:layout_marginTop="8dp"
28.         android:text="滚蛋吧！肿瘤君"
29.         android:textColor="#FF0000"
30.         app:layout_constraintBottom_toBottomOf="parent"
31.         app:layout_constraintEnd_toEndOf="parent"
32.         app:layout_constraintStart_toStartOf="parent"
33.         app:layout_constraintTop_toBottomOf="@+id/imageView" />
34. </android.support.constraint.ConstraintLayout>
```

第 19 行代码设置 ImageView 组件的图片源。

第 28 行代码设置文字内容。

第 29 行代码设置文字颜色。

（5）在项目结构中找到 src 目录，打开源程序 MainActivity.java，重写 onCreate()方法，代码如下。

```
1.  public class MainActivity extends AppCompatActivity {
2.      private ImageView iv;
3.      private Button bt;
4.      protected void onCreate(Bundle savedInstanceState) {
5.          super.onCreate(savedInstanceState);
6.          setContentView(R.layout.activity_main);
```

```
7.          bt=this.findViewById(R.id.button);
8.          iv=this.findViewById(R.id.imageView);
9.          bt.setOnClickListener(new View.OnClickListener() {
10.             @Override
11.             public void onClick(View view) {
12.                 iv.setVisibility(View.GONE);
13.             }
14.         });
15.     }
16. }
```

第 2 行代码声明一个 ImageView 组件。

第 3 行代码声明一个 Button 组件。

第 7~8 行代码利用 findViewById()方法在 Activity 中获取 XML 布局文件中的相应组件。

第 9 行代码为按钮增加单击监听功能。

第 12 行代码设置图片隐藏。

(6) 保存项目, 运行。

第 2 章 项目简介

教学目标

◇ 了解设计文档的基本概念。
◇ 掌握设计文档的主要组成部分。
◇ 了解"良心食品"App 的开发任务。

 2.1 工作任务概述

通过研读"良心食品"App 设计文档,分析及熟悉"良心食品"App 的开发任务。

 2.2 初识设计文档

2.2.1 设计文档概述

设计文档是技术规范,用于描述解决问题的思路,要在开始编码之前编写。设计文档是确保正确完成工作的有力工具,它一方面是为了让开发及设计人员展开缜密的思考,另一方面是为了让其他人了解整个系统(或作为参考)。设计文档包括程序系统的基本处理流程、程序系统的组织结构、模块划分、功能分配、接口设计、运行设计、安全设计、数据结构设计和出错处理设计等。

2.2.2 设计文档模板

---

1 引言

1.1 编写目的(阐明开发本软件的目的)
1.2 项目背景

标识待开发软件的名称、代码。

列出本项目的任务提出者、项目负责人、系统分析员、系统设计人员、程序设计员、程序员、资料员，以及与本项目开展工作直接有关的人员和用户。

说明该软件与其他有关软件的相互关系。

1.3 术语说明

列出本文档中所用到的专门术语的定义和英文缩写词的原文。

1.4 参考资料

2 项目概述

2.1 待开发软件的一般描述

描述待开发软件的背景、所应达到的目标及市场前景等。

2.2 待开发软件的功能

2.3 用户特征和水平

描述最终用户应具有的受教育水平、工作经验及技术专长。

2.4 运行环境

描述软件的运行环境，包括硬件平台、硬件要求、操作系统和版本，以及其他的软件或与其共存的应用程序等。

2.5 条件与限制

给出影响开发人员在设计软件时的约束条款。

3 功能需求

3.1 功能划分

列出所开发的软件能实现的全部功能，可采用文字、图表或数学公式等多种方法进行描述。

3.2 功能描述

对各个功能进行详细的描述。

4 外部接口需求

4.1 用户界面

对用户希望该软件所具有的界面特征进行描述。

4.2 硬件接口

描述系统中软件和硬件设备每个接口的特征、硬件接口支持的设备、软件与硬件接口之间及硬件接口号支持设备之间的约定，包括交流的数据和控制信息的性质及所使用的通信协议。

4.3 软件接口

描述该软件与有关软件的接口关系，并指出这些外部软件或组件的名称和版本号。

4.4 通信接口

描述该软件与有关软件的接口关系，并指出这些外部软件或组件的名称和版本号。

4.5 故障处理

对可能的软件故障、硬件故障及各项性能而言所产生的后果进行处理。

5 性能需求

5.1 数据精确度

输出结果的精度。

5.2 时间特性

时间特性可包括响应时间、更新处理时间、数据转换与传输时间、运行时间等。

5.3 适应性

在操作方式、运行环境、与其他软件的接口、开发计划等发生变化时,软件的适应能力。

6 其他需求

列出在本文的其他部分未出现的需求。

7 数据描述

7.1 静态数据

7.2 动态数据

包括输入数据和输出数据。

7.3 数据库描述

给出要使用的数据库的名称和类型。

7.4 数据字典

7.5 数据采集

列出提供输入数据的机构、设备和人员。
列出数据输入的手段、介质和设备。
列出数据生成方法、介质和设备。

8 附录

2.3 分析开发任务

1. 知识点

设计文档的主要组成部分。

2. 工作任务

通过细心研读"良心食品"App 设计文档,分析其开发任务。

3. "良心食品" App 设计文档

1 项目背景

随着移动通信方式和智能手机的蓬勃发展，我国网民的数量急剧上升。在我国，当前手机 App 多集中于购物、娱乐、社交、游戏等领域，农业方面的 App 数量较少、综合性功能较差、普及范围较小。因此，开发出一款智能手机的农产品销售 App 将有广阔的前景，它可节约商务成本，尤其可以节约商务沟通和非实物交易的成本。这个 App 还可以极大地提高商务效率，让生产者能销售，购买者能安心，通过营造一个生态链，利用互联网的互联高效实现销售与消费的共赢。

2 系统设计

2.1 系统目标

（1）以直观的、可预测的方式来设计导航。
（2）通过系列交互指导用户操作。
（3）通过产品展示方式向消费者展示商城及商品的优势。
（4）有条理地展示用户的订单详情，使用户可以便捷、有效地进行相关维权处理，同时便于查看商品交易订单的详情。
（5）添加收藏功能以便于消费者下次消费，促进二次交易。
（6）完善的系统注册功能。
（7）系统最大限度地实现易维护性和易操作性。

2.2 系统功能结构（见图 2-1）

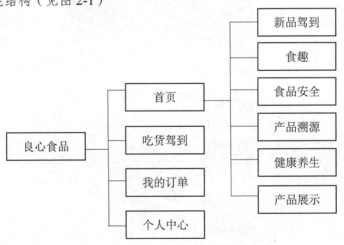

图 2-1 "良心食品" App 功能结构图

2.3 系统预览

"良心食品" App 由移动端和服务器端组合而成。下面仅列出几张典型效果图。

（1）移动端的"首页"如图 2 所示。移动端的"首页"用于功能入口的聚合展示，分别将"新品驾到""食趣""食品安全""产品溯源""健康养生""产品展示"等子功能展示出来，便于用户使用 App。

图 2-2　移动端的"首页"

（2）移动端的"吃货驾到"页面如图 2-3 所示，该页面展示了会员分享的美食信息。

图 2-3　移动端的"吃货驾到"页面

（3）移动端的"我的订单"页面如图 2-4 所示，该页面展示了用户的最近订单及收藏信息。

图 2-4 移动端的"我的订单"页面

（4）移动端的"个人中心"页面如图 2-5 所示，可以在此页面查看个人信息、修改密码、查看收藏信息等。

图 2-5 移动端的"个人中心"页面

（5）移动端的"登录"页面如图 2-6 所示，该页面主要用于用户的登录验证。

图 2-6 移动端的"登录"页面

3 数据库逻辑结构设计

根据设计好的 E-R 图在数据库中创建数据表,本教学项目只需要两个数据表:一个是分享信息表,用于用户分享美食信息,保存于本地;另一个是用户信息表,用于存储用户信息,保存于服务器端。

表 2-1 分享信息表

字段	类型	中文含义	备注说明
_id	integer	ID	primary key autoincrement
name	text	发布者名称	
date	text	发布时间	
comment	text	发布内容	
image	BLOB	发布图片	

表 2-2 用户信息表

字段	类型	中文含义	备注说明
userid	varchar(15)	用户ID	邮箱或电话号码primary key
username	varchar(50)	用户名	
password	varchar(15)	用户密码	

4 文件夹组织结构

每个项目都会有相应的文件夹组织结构,如果项目中的文件数量较多,可以将不同类型的文件存放于不同的文件夹中。"良心食品"项目的文件夹组织结构如图 7 所示。

图2-7 "良心食品"项目的文件夹组织结构

5 软件开发需求

运行环境：Windows 操作系统、Android 智能手机等。
开发语言：Java。
开发软件：Android Studio。
开发插件：JDK、SDK。

第 3 章 "登录"模块的布局

  教学目标

✧ 了解 Activity 对应的 UI 布局创建过程。
✧ 掌握 LinearLayout 的常用属性及其使用方法。
✧ 掌握 EditText 的常用属性及其使用方法。
✧ 掌握 Android 图片不同分辨率的适配。
✧ 掌握 res/values 文件夹下各类资源文件的使用方法。
✧ 掌握 shape 的使用方法。
✧ 掌握 selector 的使用方法。

 ## 3.1 工作任务概述

本章工作任务主要完成"良心食品"App 登录界面的 UI 布局,具体 UI 效果如图 3-1 所示。

图 3-1 "良心食品"App 登录界面

3.2 预备知识

3.2.1 View 与 ViewGroup 布局

Android 的 UI 界面都是由 View 和 ViewGroup 及其派生类组合而成的。其中，View 是所有 UI 组件的基类；而 ViewGroup 是容纳这些 UI 组件的容器，其本身也是由 View 派生的。

View 是 Android 平台用户 UI 界面的最基础单元。View 类为其 widgets（工具）子类奠定了基础。View 组件是可见的视觉组件，在其内部不能再置入其他组件。

ViewGroup 类为其 Layouts（布局）子类奠定了基础。ViewGroup 组件是不可见的容器组件，用来设置其容器的 View 组件和 ViewGroup 组件的排列规则。View 与 ViewGroup 的关系如图 3-2 所示，一般在 Android Studio 中可以通过 Component Tree 视图查看它们之间的树形结构。

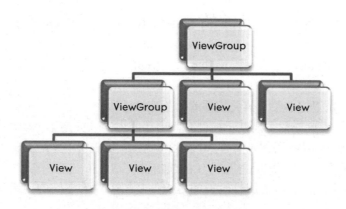

图 3-2　View 与 ViewGroup 的关系

3.2.2 LinearLayout

1. LinearLayout 简介

LinearLayout 是线性布局控件，其包含的子控件以横向或竖向的方式排列，并按照相对位置来排列所有的 widgets 或其他的 containers，当超过边界时，某些控件将缺失或消失。因此，一个垂直列表的每一行只有一个 widget 或 container（两者的宽度不限）；一个水平列表只有一个行高（高度为最高子控件的高度加上边框高度）。LinearLayout 保持其所包含的 widget 或 container 之间的间隔及相互对齐方式（相对于一个控件的右对齐、中间对齐或左对齐）。

2. 线性布局常用属性及其作用（见表 3-1）

表 3-1　线性布局常用属性及其作用

属　　性	作　　用
android:contentDescription	定义简要描述视图内容的文本

续表

属性	作用
android:layout_width; android:layout_height	这两个属性可以简单理解为View的宽与高，它们的值选项中的match_parent、wrap_content、fill_parent代表此View在父View中的宽与高的确定方式。match_parent、fill_parent代表此View的宽（或高）和父View的宽（或高）相等，wrap_content代表此View的宽高值会按照包裹自身内容的方式来确定
android:orientation	设置其内容的对齐方向（vertical表示垂直线性布局，horizontal表示水平线性布局）
android:gravity	指定该对象中放置内容的对齐方式
android:layout_gravity	相对于它的父元素的对齐方式
android:layout_weight	通过设置控件的layout_weight属性控制各个控件在布局中的相对大小。线性布局会根据该控件的layout_weight属性值与其所处布局中所有控件的layout_weight属性值之和的比值为该控件分配占用的区域

3.2.3 Android 中控件的 margin 属性和 padding 属性

Android 属性中的 margin 属性和 padding 属性是布局中比较常用的两个属性，这两个属性主要用来设置边距。

margin 属性：设置控件距离其父控件或兄弟控件的边距。

padding 属性：设置控件距离其子控件或其内部的内容（如文本）的边距。

margin 属性与 padding 属性示意图如图 3-3 所示。若以控件 B 为主，设置控件 B 的 margin 属性和 padding 属性，则控件 A 是控件 B 的父控件，控件 C（或内容 C）是控件 B 的子控件或内部内容。控件 B 的 margin 属性设置的是控件 B 与控件 A 之间的距离；控件 B 的 padding 属性设置的是控件 B 与控件 C（或内容 C）之间的距离。

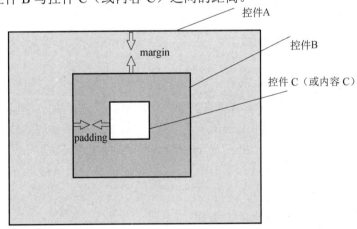

图 3-3　margin 属性与 padding 属性示意图

3.2.4 EditText 组件

在 Android 中，EditText（编辑框）用于在屏幕上显示文本输入框，在其中可以输入单行文本，也可以输入多行文本，还可以输入指定格式的文本（如密码、电话号码和邮箱等）。EditText 组件的常用属性及其作用如表 3-2 所示。

表 3-2　EditText 组件的常用属性及其作用

属　　性	作　　用
android:hint	设置显示在编辑框中的提示信息
android:numeric	设置编辑框中输入的数据类型：integer（正整数）、signed（带符号整数，有正负之分）和 decimal（浮点数）
android:singleLine	设置是否单行输入，一旦设置为 true，则文字不会自动换行
android:password	设置文本是否以密码形式显示
android:textColor	设置文字颜色
android:textStyle	设置文字样式：bold、italic、bolditalic
android:textSize	设置文字大小
android:textColorHighlight	设置被选中文字的底色，默认为蓝色
android:textColorHint	设置提示信息文字的颜色，默认为灰色
android:textScaleX	设置字间距
android:typeface	设置字型：normal、sans、serif、monospace
android:background	设置背景
android:layout_weight	设置权重
android:drawableBottom	在文字的下方输出一个 drawable，如图片
android:drawableLeft	在文字的左边输出一个 drawable，如图片
android:drawableRight	在文字的右边输出一个 drawable，如图片
android:drawableTop	在文字的上方输出一个 drawable，如图片
android:drawablePadding	设置 text 与 drawable 的间隔。该属性与 drawableLeft、drawableRight、drawableTop、drawableBottom 结合使用可设置为负数，单独使用时没有效果
android:editable	设置是否可编辑
Android:maxlength	设置编辑框的最大可输入字符数

3.2.5　Android 图片不同分辨率的适配

1. 尺寸概念

（1）px（pixels）：像素，屏幕上的点，不同设备显示效果相同。例如，HVGA 表示 320px×480px。

（2）in：英寸，屏幕的物理尺寸，1in=2.54cm。例如，手机的屏幕大小为 5in、4in，这些尺寸是屏幕的对角线长度，如 4in 表示手机屏幕（可视区域）的对角线长度是 4×2.54=10.16cm。

（3）pt（point）：标准长度单位，1pt=1/72in，用于印刷业，iOS 字体单位，Android 项目开发不涉及。

（4）dpi（dots per inch）：打印分辨率，每英寸所能打印的点数（每英寸包含的像素数），即打印精度。例如，对于分辨率为 320px×480px、宽为 2in、高为 3in 的手机，其屏幕每英寸包含的像素的数量为 320/2=160dpi（横向）或 480/3=160dpi（纵向），160 就是这部手机的 dpi，横向和纵向的这个值都是相同的（因为大部分手机屏幕使用正方形的像素点）。

（5）ppi（pixels per inch）：图像分辨率、像素密度，指图像每英寸所包含的像素数。

（6）density：屏幕密度。density 和 dpi 的关系为 1density=dpi/160。

（7）dp（dip, device independent pixels）：设备独立像素，是 Android 特有的单位，与密度无关，是基于屏幕密度的抽象单位。对于分辨率为 320px×480px 同时 dpi 为 160 的显示器，1dp=1px。

（8）sp（scaled pixels）：放大像素，与刻度无关，是文字大小单位，可以根据用户的文字大小首选项进行缩放。sp 也是 Android 特有的单位。由 TextView 的源码可知，Android 默认使用 sp 作为文字大小单位。以 160ppi 屏幕为标准，当字体大小为 100%时，1sp=1px。

2. 换算关系

（1）px = dp×（dpi / 160）。

 小贴士

用 sp 和 dp 代替 px 的原因是它们不会随 ppi 的变化而变化，在物理尺寸相同、ppi/dpi 不同的情况下，它们呈现的高度是相同的，也就是说，sp 和 dp 更接近物理呈现，而 px 则不行。

（2）$ppi=\sqrt{(长度像素数^2+宽度像素数^2)}$/屏幕对角线英寸数。

3. 区分标准

Google 官方指定的 dpi 区分标准如表 3-3 所示。

表 3-3　Google 官方指定的 dpi 区分标准

名称	ppi范围	dpi范围	图片icon尺寸
drawable-ldpi	120～160	0～120	36×36
drawable-mdpi	160～240	120～160	48×48
drawable-hdpi	240～320	160～240	72×72
drawable-xhdpi	320～480	240～320	96×96
drawable-xxhdpi	480～640	320～480	144×144
drawable-xxxhdpi	640～800	480～640	192×192

下面举例说明 Android 手机如何找到与之适配的图片。例如，某款手机配置为 1080px×1920px 和 400dpi，则对应 drawable-xxhdpi 文件夹，Android 会自动优先在 drawable-xxhdpi 文

件夹中寻找对应的图片。如果找到对应图片则加载，此时图片在手机屏幕上显示的就是其本身的大小；如果未找到，Android 会到更高分辨率的 drawable-xxxhdpi 文件夹中寻找，若一直寻找到最高分辨率的文件夹也没有的话，就开始由高到低依次查找低分辨率的文件夹，即从 drawable- xhdpi 文件夹一直查找到 drawable-ldpi 文件夹。

3.2.6　res/values 文件夹下常用的 XML 资源文件

在所有 XML 资源文件的目录设置中，最常使用的文件夹是 res\values，在此文件夹中一般会创建 string.xml、color.xml、dimens.xml、styles.xml 四种类型的 XML 资源文件。

1. string.xml（文字资源文件）

为了体现国际化及减小 App 的体积，降低数据的冗余，在 Android 开发中会把应用程序中出现的文字单独存放在 string.xml 中。作为 Android 应用开发人员，一定要养成良好的编程习惯。

（1）在 string.xml 文件中添加字符串，具体代码如下。

```
1. <?xml version="1.0" encoding="utf-8"? >
2. <resources>
3.     <string name="hello">Hello World，MainActivity！</string>
4.     <string name="app_name">TestExample01</string>
5. </resources>
```

（2）在 Java 源代码中使用 getString(R.string.app_name)。
（3）在 UI 布局文件中使用 android:text="@string/ app_name"。

2. colors.xml（颜色资源文件）

color.xml 文件中主要设置应用程序中所需的颜色。Android 的文字颜色定义方式采用类似网页格式的颜色定义方式，即常见的十六进制法。颜色设置语法表如表 3-4 所示。

表 3-4　颜色设置语法表

颜 色 语 法	语 法 帮 助	示范（采用十六进制）	颜　　色
#RGB	无Alpha，8位表示法	#00f	蓝色
#ARGB	有Alpha，8位表示法	#800f	半透明蓝色
#RRGGBB	无Alpha，16位表示法	#0000ff	蓝色
#AARRGGBB	有Alpha，16位表示法	#800000ff	半透明蓝色

（1）在 color.xml 文件中添加颜色配置信息，具体代码如下。

```
1. <?xml version="1.0" encoding="utf-8"?>
2. <resources>
3.     <drawablename="red">#f00</drawable>
4.     <drawablename="green">#0f0</drawable>
5.     <drawablename="gray">#ccc</drawable>
6. </resources>
```

（2）在 Java 源代码中使用 getResources().getColor(R.color.green)。

（3）在 UI 布局文件中使用 android:textColor="@color/greeen"。

3. dimens.xml（尺寸资源文件）

dimens.xml 可用于设置组件的大小及文字大小，它提供了如表 3-5 所示的几种尺寸定义方式。

表 3-5　尺寸定义表

尺寸格式	帮　助	描　述
px	pixel	以像素为单位
in	inches	以英寸为单位
mm	millimeter	以毫米为单位
pt	points	1pt=1/72英寸
dp或dip	density independent pixels	1dp=1/60英寸
sp	scale pixels	通常用于指定字体的大小，当用户修改手机显示的字体时，字体大小会随之改变

（1）在 dimens.xml 中添加尺寸配置信息，具体代码如下。

```
1.  <?xml version="1.0" encoding="utf-8"?>
2.  <resources>
3.      <dimen name="btn_width">30mm</dimen>
4.  </resources>
```

（2）在 Java 源代码中使用 getResources().getDimension(R.dimen. btn_width)。

（3）在 UI 布局文件中使用 android:layout_width="@dimen/ btn_width"。

4. styles.xml（主题风格资源文件）

styles.xml 类似于网站的样式表文件，属于更高级的 XML 资源文件，它是一个多属性的 XML 资源文件。在 Android Studio 中，styles.xml 文件会默认产生一个名字为 AppTheme 的样式，该样式是项目程序的主题样式。

（1）在 styles.xml 中添加样式信息，具体代码如下。

```
1.  <resources>
2.      <style name="text_font">
3.          <item name="android:textColor">#05b</item>
4.          <item name="android:textSize">18sp</item>
5.          <item name="android:textStyle">bold</item>
6.      </style>
7.  </resources>
```

（2）在 Java 源代码中使用 setTheme(R.style. text_font)。

（3）在 UI 布局文件中使用 style="@style/ text_font"。

3.2.7　shape

1. shape 简介

shape 是用于定义一些形状的样式，通常用于在 Android 开发中控制控件的背景。shape 共有 6 个属性，分别是 corners、padding、size、solid、stroke、gradient。

2. 在 Android Studio 中添加 shape 的方法

（1）在 Project 视图中右击 res 文件夹，依次单击"New"→"Android Resource File"选项，新建文件，如图 3-4 所示。

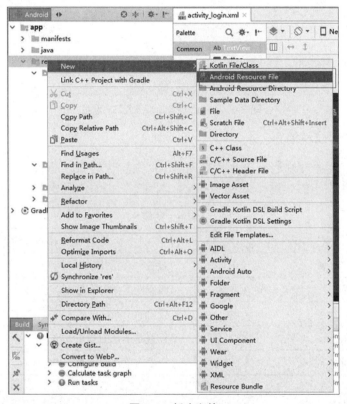

图 3-4　新建文件

（2）在打开的"New Resource File"对话框中，除了要通过"File Name"为新文件命名，还需要将"Resource type"修改为"Drawable"，并将"Root element"修改为"shape"，如图 3-5 所示。

（3）单击 OK 按钮，在项目的 res/drawable 文件夹中添加一个名为 test.xml 的 shape 文件。

（4）打开 test.xml 文件，在该文件内添加相应的属性即可。

　小贴士

shape 文件存放于 drawable 文件夹中，可以把 shape 文件看成图片，在实际应用中以图片的方式应用即可。

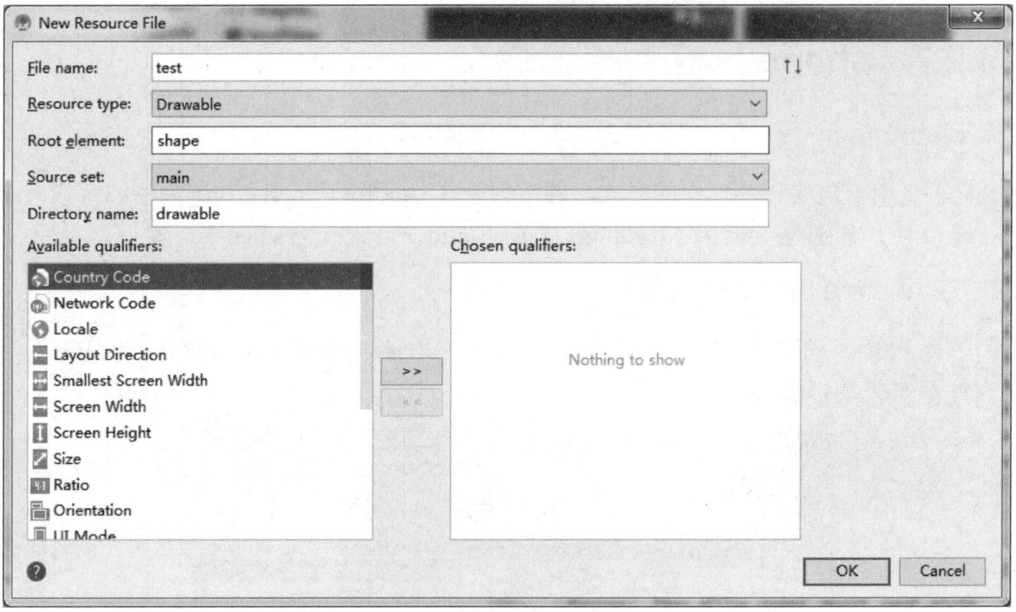

图 3-5　New Resource File 对话框

3. shape 常用属性简介

（1）corners：用于控制边框 4 个角的大小，如果默认是 0dp 的话就是直角，如果设置值大于 0dp 就会产生圆角的效果。corners 有 5 个属性，它们的作用如表 3-6 所示。

表 3-6　corners 的属性的作用

属　　性	作　　用
android:radius	设置4个角的圆角大小
android:topLeftRadius	设置左上角的圆角大小
android:topRightRadius	设置右上角的圆角大小
android:bottomLeftRadius	设置左下角的圆角大小
android:bottomRightRadius	设置右下角的圆角大小

corners 案例 1：

```
<corners android:radius="10dp"/>
```

corners 案例 1 效果如图 3-6 所示。

图 3-6　corners 案例 1 效果

corners 案例 2：

```
1.  <corners
2.      android:bottomRightRadius="10dp"
```

```
3.      android:topLeftRadius="10dp"/>
```

corners 案例 2 效果如图 3-7 所示。

图 3-7　corners 案例 2 效果

（2）padding：用于控制背景边框与背景中内容的距离，即用于控制内边距。padding 有 4 个属性，它们的作用如表 3-7 所示。

表 3-7　padding 的属性的作用

属　　性	作　　用
android:left	设置左内边距
android:right	设置右内边距
android:top	设置上内边距
android:bottom	设置下内边距

padding 案例：

```
1.  <padding
2.      android:top="20dp"
3.      android:bottom="20dp"
4.      android:left="40dp"
5.      android:right="40dp"/>
```

添加 padding 前的效果如图 3-8 所示；添加 padding 后的效果如图 3-9 所示。

图 3-8　添加 padding 前的效果　　　　图 3-9　添加 padding 后的效果

（3）size：用于设置背景的大小，有 android:height 和 android:width 两个属性，这两个属性不能设置为 match_parent 或 wrap_content，只能设置为具体的数值。另外，如果这两个属性设置得过小以至于比背景上的控件还要小的时候，系统不会以设置的数值为准，而会以包裹住控件的最小的高和宽来作为背景的高和宽；如果设置得较大，则会以设置的数值为准。size 的两个属性的作用如下。

① android:width：用于设置背景的宽度。
② android:height：用于设置背景的高度。

size 案例 1：

```
1.  <size
2.      android:height="200dp"
3.      android:width="200dp"/>
```

size 案例 1 效果如图 3-10 所示。

图 3-10　size 案例 1 效果

size 案例 2：

```
1.    <size
2.        android:height="2dp"
3.        android:width="2dp"/>
```

size 案例 2 效果如图 3-11 所示。

（4）solid：用于控制背景颜色，它只有一个 android:color 属性。

solid 案例：

```
1.    <solid
2.        android:color="#FF4081"/>
```

soild 案例效果如图 3-12 所示。

图 3-11　size 案例 2 效果　　　　　　　　图 3-12　solid 案例效果

（5）stroke：用于控制背景的边框。stroke 共有 4 个属性，它们的作用如表 3-8 所示。

表 3-8　stroke 的属性的作用

属　　性	作　　用
android:width	用于控制边框的宽度
android:color	用于控制边框的颜色
android:dashWidth	用于控制虚线段的长度
android:dashGap	用于控制虚线之间的距离

注意观察 android:dashGap 和 android:dashWidth 控制的边框是否为虚线，如果这两个属性同时设置为正数，那么边框就是虚线，这两个属性只要有一个没有设置（或被设置）为 0dp，那边框就是实线。

stroke 案例：

```
1.    <corners
2.        android:radius="50dp"/>
3.    <size
4.        android:height="100dp"
```

```
5.         android:width="100dp"/>
6.     <solid
7.         android:color="#FF4081"/>
8.     <stroke
9.         android:width="5dp"
10.        android:color="#3F51B5"
11.        android:dashWidth="20dp"
12.        android:dashGap="10dp"/>
```

stroke 案例效果如图 3-13 所示。

图 3-13 stroke 案例效果

（6）gradient：用于设置背景色的效果，一旦设置了这个属性，solid 中设置的背景颜色就不再生效。gradient 有 9 个属性，它们的作用如表 3-9 所示。当 android:type 的值不同时，有些属性不生效。

表 3-9 gradient 的属性的作用

属 性	作 用
android:type	指定渐变的类型，共有3种类型：linear（线性、默认）、radial（从中间往外扩散）、sweep（旋转扫一周）
android:startColor	设置渐变开始的颜色
android:endColor	设置渐变结束的颜色
android:centerColor	设置渐变中间的颜色
android:Angle	设置线性渐变的方向，默认从左向右。如果需要设置的话，则值应为整数并且要能被45整除，否则会报错。正数是逆时针，负数是顺时针。也就是说，angle=90是从下往上渐变，angle=-90是从上往下渐变，其他渐变无效果
android:centerX	在渐变类型为radial时，用于控制渐变圆圈中心点与左边框的距离
android:centerY	在渐变类型为radial时，用于控制渐变圆圈中心点与上边框的距离
android:gradientRadius	只有在渐变类型为radial时才生效，用于控制渐变圆圈的大小。在渐变类型为radial时，这个属性必须设置，否则会报错
android:useLevel	这个属性有两个值：true和false。当设置为true时，对应shape文件会被作为LevelListDrawable处理；一般设置为false。默认设置为false

gradient 案例：

```
1.  <gradient
2.      android:startColor="@color/white"
```

3. android:endColor="@color/black"
4. android:centerColor="#3F51B5"
5. android:type="linear"
6. android:useLevel="false"/>

gradient 案例效果如图 3-14 所示。

图 3-14　gradient 案例效果

3.2.8　selector

1. selector 简介

selector（选择器）在 Android 中通常用作组件的背景，这样可以省去用代码实现组件在不同状态下的不同背景颜色或不同图片的变换，使用十分方便。

2. 在 Android Studio 中添加 selector 的方法

（1）在 Project 视图中右击 res 文件夹，依次单击"New"→"Android Resource File"选项，新建文件，如图 3-15 所示。

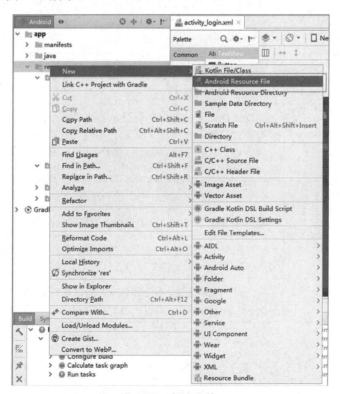

图 3-15　新建文件

（2）在打开的"New Resource File"对话框中，除了要通过"File Name"为新文件命名，还需要将"Resource type"修改为"Drawable"，将"Root element"修改为"selector"，如图 3-16 所示。

图 3-16　"New Resource File"对话框

（3）单击"OK"按钮，在项目程序的 res/drawable 文件夹中添加一个名为 test.xml 的 selector 文件。

（4）打开 test.xml 文件，在该文件中添加相应的属性即可。

3. selector 常用属性（见表 3-10）

表 3-10　selector 的属性的作用

属　性	作　用
android:color="hex_color"	设置颜色值：#RGB、$ARGB、#RRGGBB、#AARRGGBB
android:drawable="@［package:］drawable/drawable_resource"	设置图片资源
android:state_pressed=［"true" | "false"］	设置是否触摸
android:state_focused=［"true" | "false"］	设置是否获取焦点
android:state_hovered=［"true" | "false"］	设置光标是否经过
android:state_selected=［"true" | "false"］	设置是否选中
android:state_checkable=［"true" | "false"］	设置是否可勾选
android:state_checked=［"true" | "false"］	设置是否勾选
android:state_enabled=［"true" | "false"］	设置是否可用

属　性	作　用
android:state_activated=［"true"	"false"］	设置是否激活
android:state_window_focused=［"true"|"false"］	设置所在窗口是否获取焦点

4. 使用 selector 的方法

方法一：在组件中配置 android:listSelector="@drawable/xxx"，或者添加属性 android:background="@drawable/xxx"。

方法二：Drawable drawable = getResources().getDrawable(R.drawable.xxx);
组件.setSelector(drawable);

3.3　热 身 任 务

本节热身任务为微信中的"我"。

1. 任务说明

完成如图 3-17 所示的布局效果。

2. 操作步骤

（1）新建项目。

（2）将项目图片复制到项目的 res/mipmap 文件夹中。在 weixin.xml 布局文件中添加组件，产生如图 3-18 所示的效果。微信中的"我"的 Component Tree 如图 3-19 所示。

图 3-17　微信中的"我"效果图

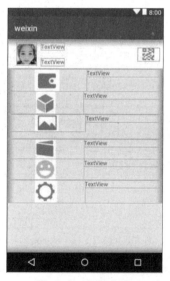

图 3-18　初始布局

图 3-19 微信中的"我"的 Component Tree

（3）修改 activity_main.xml 文件中相应组件的属性，修改完成后的 activity_main.xml 文件的代码及相应功能说明如下。

```
1.   <?xml version="1.0" encoding="utf-8"?>
2.   <LinearLayout xmlns:android="http://schemas.android.com/apk/res/android"
3.       xmlns:app="http://schemas.android.com/apk/res-auto"
4.       xmlns:tools="http://schemas.android.com/tools"
5.       android:id="@+id/linearLayout"
6.       android:layout_width="match_parent"//设置组件的宽度
7.       android:layout_height="match_parent"//设置组件的高度
8.       android:orientation="vertical"//设置线性布局为垂直线性布局
9.       android:background="#ebebeb"//设置背景颜色
10.      tools:context=".MainActivity">
11.      <android.support.constraint.ConstraintLayout
12.          android:layout_width="match_parent"
13.          android:layout_height="71dp"
14.          android:background="#ffffff">
15.          <ImageView
```

```
16.            android:id="@+id/imageView2"
17.            android:layout_width="49dp"
18.            android:layout_height="55dp"
19.            android:layout_marginBottom="8dp"//设置组件与下边组件的距离
20.            android:layout_marginEnd="8dp"//设置组件与右边组件的距离
21.            android:layout_marginStart="15dp"//设置组件与左边组件的距离
22.            android:layout_marginTop="8dp" //设置组件与上边组件的距离
23.            app:layout_constraintBottom_toBottomOf="parent"//设置与父窗体底端对齐
24.            app:layout_constraintEnd_toEndOf="parent"
25.            app:layout_constraintHorizontal_bias="0.024"//水平方向偏移量
26.            app:layout_constraintStart_toStartOf="parent"
27.            app:layout_constraintTop_toTopOf="parent"
28.            app:srcCompat="@mipmap/userphoto" />//设置 ImageView 的图像源为 userphoto
29.        <TextView
30.            android:id="@+id/textView"
31.            android:layout_width="wrap_content"
32.            android:layout_height="wrap_content"
33.            android:layout_marginLeft="10dp"
34.            android:text="妲己"//设置文本内容为"妲己"
35.            app:layout_constraintLeft_toRightOf="@id/imageView2"
36.            app:layout_constraintTop_toTopOf="@id/imageView2" />
37.        <TextView
38.            android:id="@+id/textView2"
39.            android:layout_width="wrap_content"
40.            android:layout_height="wrap_content"
41.            android:layout_marginLeft="10dp"
42.            android:text="微信号：oh my god"
43.            app:layout_constraintBottom_toBottomOf="@id/imageView2"
44.            app:layout_constraintLeft_toRightOf="@id/imageView2"/>
45.        <ImageView
46.            android:id="@+id/imageView3"
47.            android:layout_width="56dp"
48.            android:layout_height="32dp"
49.            android:layout_marginBottom="8dp"
50.            android:layout_marginEnd="8dp"
51.            android:layout_marginStart="8dp"
52.            android:layout_marginTop="8dp"
53.            app:layout_constraintBottom_toBottomOf="parent"
54.            app:layout_constraintEnd_toEndOf="parent"
55.            app:layout_constraintHorizontal_bias="1.0"
56.            app:layout_constraintStart_toStartOf="parent"
57.            app:layout_constraintTop_toTopOf="parent"
58.            app:layout_constraintVertical_bias="0.478"//垂直方向偏移量
59.            app:srcCompat="@mipmap/erwanma" />
60.    </android.support.constraint.ConstraintLayout>
61.    <LinearLayout
62.        android:layout_width="match_parent"
```

```
63.        android:layout_height="40dp"
64.        android:layout_marginTop="30dp"
65.        android:background="#ffffff"
66.        android:gravity="center_vertical"//设置组件内部元素垂直居中对齐
67.        android:orientation="horizontal">
68.        <ImageView
69.            android:id="@+id/imageView4"
70.            android:layout_width="20dp"
71.            android:layout_height="20dp"
72.            android:layout_weight="1"
73.            app:srcCompat="@mipmap/icon1" />
74.        <TextView
75.            android:id="@+id/textView3"
76.            android:layout_width="wrap_content"
77.            android:layout_height="wrap_content"
78.            android:layout_weight="4"
79.            android:text="钱包" />
80.    </LinearLayout>
81.    <LinearLayout
82.        android:layout_width="match_parent"
83.        android:layout_height="40dp"
84.        android:layout_marginTop="30dp"
85.        android:background="#ffffff"
86.        android:gravity="center_vertical"
87.        android:orientation="horizontal">
88.        <ImageView
89.            android:id="@+id/imageView5"
90.            android:layout_width="20dp"
91.            android:layout_height="20dp"
92.            android:layout_weight="1"
93.            app:srcCompat="@mipmap/icon2" />
94.        <TextView
95.            android:id="@+id/textView4"
96.            android:layout_width="wrap_content"
97.            android:layout_height="wrap_content"
98.            android:layout_weight="4"
99.            android:text="收藏 " />
100.   </LinearLayout>
101.   <LinearLayout
102.       android:layout_width="match_parent"
103.       android:layout_height="40dp"
104.       android:layout_marginTop="2dp"
105.       android:background="#ffffff"
106.       android:gravity="center_vertical"
107.       android:orientation="horizontal">
108.       <ImageView
109.           android:id="@+id/imageView6"
```

```
110.            android:layout_width="20dp"
111.            android:layout_height="20dp"
112.            android:layout_weight="1"
113.            app:srcCompat="@mipmap/icon3" />
114.        <TextView
115.            android:id="@+id/textView5"
116.            android:layout_width="wrap_content"
117.            android:layout_height="wrap_content"
118.            android:layout_weight="4"
119.            android:text="相册" />
120.    </LinearLayout>
121.    <LinearLayout
122.        android:layout_width="match_parent"
123.        android:layout_height="40dp"
124.        android:layout_marginTop="2dp"
125.        android:background="#ffffff"
126.        android:gravity="center_vertical"
127.        android:orientation="horizontal">
128.        <ImageView
129.            android:id="@+id/imageView7"
130.            android:layout_width="20dp"
131.            android:layout_height="20dp"
132.            android:layout_weight="1"
133.            app:srcCompat="@mipmap/icon4" />
134.        <TextView
135.            android:id="@+id/textView6"
136.            android:layout_width="wrap_content"
137.            android:layout_height="wrap_content"
138.            android:layout_weight="4"
139.            android:text="卡包" />
140.    </LinearLayout>
141.    <LinearLayout
142.        android:layout_width="match_parent"
143.        android:layout_height="40dp"
144.        android:layout_marginTop="2dp"
145.        android:background="#ffffff"
146.        android:gravity="center_vertical"
147.        android:orientation="horizontal">
148.        <ImageView
149.            android:id="@+id/imageView8"
150.            android:layout_width="20dp"
151.            android:layout_height="20dp"
152.            android:layout_weight="1"
153.            app:srcCompat="@mipmap/icon5" />
154.        <TextView
155.            android:id="@+id/textView7"
156.            android:layout_width="wrap_content"
```

```
157.            android:layout_height="wrap_content"
158.            android:layout_weight="4"
159.            android:text="表情" />
160.     </LinearLayout>
161.     <LinearLayout
162.         android:layout_width="match_parent"
163.         android:layout_height="40dp"
164.         android:layout_marginTop="30dp"
165.         android:background="#ffffff"
166.         android:gravity="center_vertical"
167.         android:orientation="horizontal">
168.         <ImageView
169.            android:id="@+id/imageView9"
170.            android:layout_width="20dp"
171.            android:layout_height="20dp"
172.            android:layout_weight="1"
173.            app:srcCompat="@mipmap/icon6" />
174.         <TextView
175.            android:id="@+id/textView8"
176.            android:layout_width="wrap_content"
177.            android:layout_height="wrap_content"
178.            android:layout_weight="4"
179.            android:text="设置" />
180.     </LinearLayout>
181. </LinearLayout>
```

思考

还有其他方法可以实现与微信中的"我"相同的布局吗？

上述方法中的代码还可以简化吗？

3.4 实现"登录"模块的布局

1. 知识点

➢ LinearLayout 的使用方法。
➢ Android 图片不同分辨率的适配。
➢ EditText 的使用方法。
➢ shape 的使用方法。
➢ selector 的使用方法。
➢ res/values 文件夹下各类资源文件的使用方法。

2. 工作任务

制作"良心食品"App 登录模块的 UI 布局，实现如图 3-20 所示的 UI 效果。

图 3-20 "登录"界面

3. 操作流程

(1) 新建项目,项目名称为 HealthFood。

(2) 将项目用到的所有图片分类复制到 Drawable 和 mipmap 两个文件夹中(注意要根据 Android 图片不同分辨率适配原则分别存放图片)。

(3) 右击项目中的 res/layout/activity_main.xml 文件,依次单击"refactor"→"Rename"选项,将文件名改为 activity_login.xml。

(4) 在 activity_login.xml 文件中添加组件,为了避免在一个项目中出现 ID 重名问题,建议添加组件时在每个组件的默认 ID 名前面都添加一个"Login_",完成后的登录界面 UI 效果图如图 3-21 所示,其 Component Tree 如图 3-22 所示。

图 3-21 完成后的"登录"界面 UI 效果图

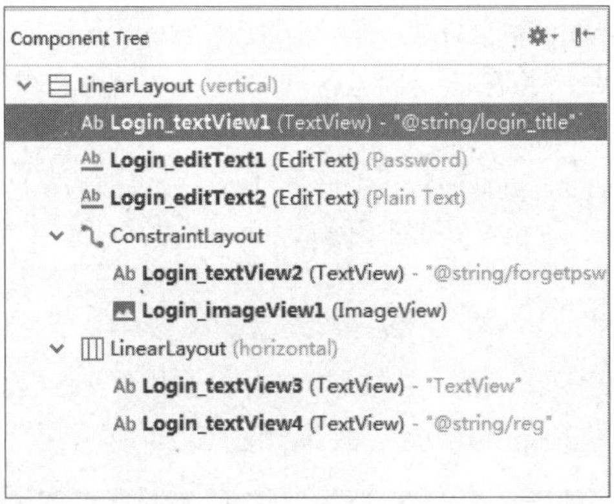

图 3-22 "登录"界面的 Component Tree

（5）选中 Component Tree 中最外层的 LinearLayout 组件，通过修改其 background 属性将图片 login_bg.jpg 设置为整个布局的背景。

（6）打开项目 res/values 文件夹下的 strings.xml 文件，修改该文件的 app_name 属性，并添加 login_title 等多个字符串属性，具体代码如下。

```
1.  <?xml version="1.0" encoding="utf-8"?>
2.  <resources>
3.      <string name="app_name">良心食品</string>//
4.      <string name="login_title">登入</string>
5.      <string name="username">手机号/邮箱</string>
6.      <string name="password">密 码</string>
7.      <string name="forgetpsw">忘记密码？点击找回</string>
8.      <string name="nouser">还没有账号？</string>
9.      <string name="reg">注册</string>
10. </resources>
```

第 3 行代码用于修改 App 标题。该行代码默认通过 AndroidManifest.xml 文件中的 android:label="@string/app_name"语句实现对 App 标题的修改。

（7）通过属性面板将组件的 text 属性设置为在步骤（6）中添加的字符串。例如，将 Login_textView1 组件的 text 属性设置为@string/login_title，完成修改后的布局效果如图 3-23 所示。

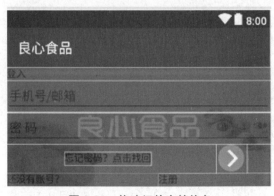

图 3-23 修改组件字符信息

（8）右击项目中的 res/values 文件夹，依次单击"New"→"XML"→"Values XML File"选项，新建 dimens.xml 文件，并在该文件中添加 login_textsize 属性，具体代码如下。

```
1.    <resources>
2.        <!-- Default screen margins, per the Android Design guidelines. -->
3.        <dimen name="login_textsize">30sp</dimen>//此尺寸定义用于设置"登入"的字体大小
4.        <dimen name="edit_dimens">10dp</dimen>//此尺寸定义用于设置文本编辑框圆角的大小
5.    </resources>
```

（9）通过属性面板将在步骤（8）中添加的尺寸应用于 TextView 的 TextSize 属性，实现将"登入"文字大小修改为 30sp 的功能，效果如图 3-24 所示。

图 3-24　修改组件尺寸信息

（10）打开项目 res/values 文件夹下的 colors.xml 文件，并在该文件中添加 edit_color 等属性，具体代码如下。

```
1.    <?xml version="1.0" encoding="utf-8"?>
2.    <resources>
3.        <color name="edit_color">#40ffffff</color>//此颜色定义用于设置编辑框中文字的颜色
4.        <color name="hint_textColor">#65ffffff</color>//此颜色定义用于设置编辑框中提示文字的颜色
5.        <color name="font_color">#ffffff</color>//此颜色定义用于设置白色文字
6.    </resources>
```

（11）在 Project 视图中右击 src 文件夹，依次单击"New"→"Android Resource File"选项，新建一个名字为 edit_login_shape_t.xml 的 shape 文件，并在该文件中添加相应属性。此处的 shape 文件用于设置 Login_editText1 及 Login_editText2 两个文本编辑框获取焦点后的背景效果，具体代码如下。

```
1.    <?xml version="1.0" encoding="utf-8"?>
2.    <shape xmlns:android="http://schemas.android.com/apk/res/android" >
3.        //设置 shape 圆角为 Dimens 中定义的 edit_dimens
4.        <corners android:radius="@dimen/edit_dimens"></corners>
5.        <solid android:color="@color/hint_textColor"/>//设置 shape 描边颜色为 Colors 中定义的 hint_textColor
6.    </shape>
```

（12）重复步骤（11）的操作，新建一个名字为 edit_login_shape_f.xml 的 shape 文件，并在该文件中添加相应属性。此处的 shape 文件用于设置 Login_editText1 及 Login_editText2 两

个文本编辑框失去焦点后的背景效果,具体代码如下。

```
1.  <?xml version="1.0" encoding="utf-8"?>
2.  <shape xmlns:android="http://schemas.android.com/apk/res/android" >
3.      <corners android:radius="@dimen/edit_dimens" ></corners>
4.      <solid android:color="@color/edit_color"/>
5.  </shape>
```

(13) 在 Project 视图中右击 src 文件夹,依次单击"New"→"Android Resource File"选项,新建一个名为 edituser_selector 的 selector 文件,并在该文件中添加相应属性。此处的 selector 文件主要用于完成当 Login_editText1 及 Login_editText2 两个文本编辑框获取焦点或失去焦点时自动切换背景的功能,具体代码如下。

```
1.  <?xml version="1.0" encoding="utf-8"?>
2.  <selector xmlns:android="http://schemas.android.com/apk/res/android" >
3.      <item android:state_focused="true"
4.          //将组件获取焦点后的图片设置为 edit_login_shape_t.xml
5.          android:drawable="@drawable/edit_login_shape_t "></item>
6.      <item android:state_focused="false"
7.          //将组件失去焦点后的图片设置为 edit_login_shape_f.xml
8.          android:drawable="@drawable/edit_login_shape_f "></item>
9.  </selector>
```

(14) 打开项目 res/values 文件夹下的 style.xml 文件,并在该文件中添加 login_et 样式,具体代码如下。

```
1.  <style name="login_et">
2.      <item name="android:layout_width">match_parent</item>
3.      //设置组件的宽度与父窗体对齐
4.      <item name="android:layout_height">45dp</item>
5.      //设置组件的高度为 45dp
6.      <item name="android:textColor">#ffffff</item>
7.      //设置文字颜色为白色
8.      <item name="android:background">@drawable/edituser_selector</item>
9.      //设置背景为 edituser_selector
10.     <item name="android:layout_marginLeft">30dp</item>
11.     //设置与左边元素边缘的距离为 30dp
12.     <item name="android:layout_marginRight">30dp</item>
13.     //设置与右边元素边缘的距离为 30dp
14.     <item name="android:maxLength">20</item>
15.     //设置最多可输入 20 个字符
16.     <item name="android:textColorHint">@color/hint_textColor</item>
17.     //设置提示文字的颜色为颜色资源文件中定义的 hint_textColor
18.     <item name="android:textSize">@dimen/login_textsize</item>
19.     //设置文字大小为尺寸资源文件中定义的 login_textsize
20.     <item name="android:layout_gravity">center_horizontal</item>
21.     //定义组件相对于父窗体水平居中对齐
22.     <item name="android:drawablePadding">8dp</item>
23.     //设置 text 与 drawable(图片)的间隔为 8dp
24.     <item name="android:singleLine">true</item>
25.     //设置为单行文本显示
```

26.   </style>

（15）分别在 Login_editText1 及 Login_editText2 两个文本编辑框组件中将 style 属性设置为@style/lgoin_et，实现对这两个文本编辑框样式的添加。

（16）分别在 Login_editText1 及 Login_editText2 两个文本编辑框组件中将 drawableLeft 属性设置为@drawable/user 和@drawable/pass，实现对这两个文本编辑框左内置图的添加。

（17）分别设置其他组件的文字颜色、间距等属性，最终完成布局，具体代码如下。

```
1.  <LinearLayout xmlns:android="http://schemas.android.com/apk/res/android"
2.      xmlns:app="http://schemas.android.com/apk/res-auto"
3.      xmlns:tools="http://schemas.android.com/tools"
4.      android:layout_width="match_parent"
5.      android:layout_height="match_parent"
6.      android:background="@drawable/login_bg"
7.      android:orientation="vertical"
8.      tools:context=".Login">
9.      <TextView
10.         android:id="@+id/Login_textView1"
11.         android:layout_width="match_parent"
12.         android:layout_height="wrap_content"
13.         android:layout_marginLeft="30dp"
14.         android:layout_marginTop="160dp"
15.         android:text="@string/login_title"
16.         android:textColor="@color/font_color"
17.         android:textSize="@dimen/login_textsize" />
18.     <EditText
19.         android:id="@+id/Login_editText1"
20.         style="@style/Login_et"
21.         android:layout_marginTop="30dp"
22.         android:drawableLeft="@mipmap/user"//图片放在 mipmap 下，所以用@mipmap/方式引用
23.         android:ems="10"
24.         android:hint="@string/username"
25.         android:inputType="textPassword" />
26.     <EditText
27.         android:id="@+id/Login_editText2"
28.         style="@style/login_et"
29.         android:layout_marginTop="30dp"
30.         android:drawableLeft="@mipmap/pass"
31.         android:ems="10"
32.         android:hint="@string/password"
33.         android:inputType="textPersonName" />
34.     <android.support.constraint.ConstraintLayout
35.         android:layout_width="match_parent"
36.         android:layout_height="40dp"
37.         android:layout_marginTop="10dp">
38.         <TextView
39.             android:id="@+id/Login_textView2"
40.             android:layout_width="wrap_content"
41.             android:layout_height="wrap_content"
```

```
42.            android:layout_marginEnd="8dp"
43.            android:layout_marginStart="8dp"
44.            android:text="@string/forgetpsw"
45.            android:textColor="@color/font_color"
46.            app:layout_constraintBottom_toBottomOf="parent"
47.            app:layout_constraintEnd_toEndOf="parent"
48.            app:layout_constraintHorizontal_bias="0.316"
49.            app:layout_constraintStart_toStartOf="parent"
50.            app:layout_constraintTop_toTopOf="parent"
51.            app:layout_constraintVertical_bias="0.51" />
52.        <ImageView
53.            android:id="@+id/Login_imageView1"
54.            android:layout_width="wrap_content"
55.            android:layout_height="wrap_content"
56.            android:layout_marginBottom="8dp"
57.            android:layout_marginEnd="8dp"
58.            android:layout_marginStart="8dp"
59.            android:layout_marginTop="8dp"
60.            android:src="@mipmap/login"
61.            app:layout_constraintBottom_toBottomOf="parent"
62.            app:layout_constraintEnd_toEndOf="parent"
63.            app:layout_constraintHorizontal_bias="0.916"
64.            app:layout_constraintStart_toStartOf="parent"
65.            app:layout_constraintTop_toTopOf="parent"
66.            app:layout_constraintVertical_bias="0.523" />
67.    </android.support.constraint.ConstraintLayout>
68.    <LinearLayout
69.        android:layout_width="match_parent"
70.        android:layout_height="match_parent"
71.        android:gravity="bottom|center_horizontal"
72.        android:orientation="horizontal"
73.        android:paddingBottom="12dp">
74.        <TextView
75.            android:id="@+id/Login_textView3"
76.            android:layout_width="wrap_content"
77.            android:layout_height="wrap_content"
78.            android:text="@string/nouser"
79.            android:textColor="#bbffffff"/>
80.        <TextView
81.            android:id="@+id/Login_textView4"
82.            android:layout_width="wrap_content"
83.            android:layout_height="wrap_content"
84.            android:text="@string/reg"
85.            android:textColor="@color/font_color" />
86.    </LinearLayout>
87.</LinearLayout>
```

第 4 章 "底部导航"模块的设计

 教学目标

◇ 了解 Context。
◇ 掌握在 Android 中建立 RadioGroup 和 RadioButton 的方法。
◇ 掌握 RadioGroup 的常用属性。
◇ 掌握 RadioGroup 选中状态变换的事件（监听器）。
◇ 掌握 RadioButton 的常用属性。
◇ 掌握 Toast 的使用方法。

 ## 4.1 工作任务概述

本章工作任务是实现"良心食品"App 主界面的 UI 布局及用提示消息的方式模拟底部导航功能。当单击底部导航栏的选项卡时，会以消息框的方式显示当前功能的名字。例如，当单击底部导航的"吃货驾到"选项卡时，则有消息框显示"吃货驾到"，效果如图 4-1 所示。

图 4-1　单击底部导航的"吃货驾到"选项卡的效果

4.2 预备知识

4.2.1 Context

1. Context 的概念

Context 是指上下文或场景，比如在打电话时，场景包括电话程序对应的界面及隐藏在界面背后的数据。Context 描述的是一个应用程序环境的信息，即上下文。Context 类是一个抽象类，Android 提供了该抽象类的具体实现。利用 Context 类可以获取应用程序的资源和类，也可以实现一些应用级别的操作，如启动一个 Activity、发送广播、接收 Internet 信息等。

2. 创建 Context 实例

应用程序创建 Context 实例的情况有如下几种。

（1）在创建 Application 对象时。若整个 App 只有一个 Application 对象，则每个应用程序在第一次启动时都会先创建 Application 对象。如果对应用程序启动一个 Activity 的流程比较清楚，那么创建 Application 对象的时机为在创建 handleBindApplication()方法时，Application 位于 ActivityThread.java 类中。

（2）在创建 Service 对象时。当 startService 或 bindService 时，如果系统检测到需要创建一个新的 Service 实例，就会回调 handleCreateService()方法，以完成相关数据操作。handleCreateService()函数位于 ActivityThread.java 类中。

（3）在创建 Activity 对象时。在通过 startActivity()方法或 startActivityForResult()方法请求启动一个 Activity 时，如果系统检测到需要新建一个 Activity 对象，就会回调 handleLaunchActivity()方法，该方法继而调用 performLaunchActivity()方法，以创建一个 Activity 实例，并且回调 onCreate()方法、onStart()方法等。这些方法都位于 ActivityThread.java 类中。

3. 获取 Context 的常用方法

（1）Activity.this 的 Context（一般用法）返回当前 Activity 的上下文，它是属于 Activity 的，Activity 可以摧毁它。

（2）getApplicationContext()方法返回应用的上下文，生命周期是整个应用，应用才能摧毁该上下文。

（3）getBaseContext()方法返回由构造函数指定的或由 setBaseContext()方法设置的上下文。

（4）getActivity()方法多用于 Fragment 中。

4.2.2 RadioGroup

RadioGroup（单选按钮组）是提供 RadioButton 的容器，在该容器中添加多个 RadioButton 方可使用。如果要设置单选按钮的内容，则需要使用 RadioButton 类。RadioGroup 的常用属性及其说明如表 4-1 所示。RadioGroup 的常用方法及其作用如表 4-2 所示。

表 4-1　RadioGroup 的常用属性及其说明

属　性	说　明
android:checkedButton	子单选按钮应该在默认情况下对其所在单选按钮组进行检查的ID
android:contentDescription	定义简要描述视图内容的文本
android:orientation	单选按钮排列的方式

表 4-2　RadioGroup 的常用方法及其作用

方　法	作　用
getCheckedRadioButtonId()	获取选中按钮的ID
clearCheck()	清除选中状态
setOnCheckedChangeListener (RadioGroup.OnCheckedChangeListener listener)	当一个单选按钮组中的单选按钮选中状态发生改变时调用的回调方法
check(int id)	通过传入ID来设置该选项为选中状态

4.2.3　RadioButton

RadioButton 是指单选按钮，它有两种状态，即选中状态和未选中状态。RadioButton 允许用户从一个组中选择一个选项。RadioButton 的常用属性及其说明如表 4-3 所示。RadioButton 的常用方法及其作用如表 4-4 所示。

表 4-3　RadioButton 常用属性及其说明

属　性	说　明
drawableLeft、drawableRight、drawableTop、drawableBottom	在text的左边、右边、上边、下边输出一幅drawable图片
android:button="@null"	去除RadioButton前面的圆点
android:text	显示文本
android:checked	指定选中状态。当属性值为true时，表示选中；当属性值为false时，表示未选中。属性值默认为false

表 4-4　RadioButton 的常用方法及其作用

方　法	作　用
setCompoundDrawablesWithIntrinsicBounds()	setCompoundDrawablesWithIntrinsicBounds(Drawable top,Drawable bottom, Drawable left,Drawable right)可以在上、下、左、右设置图标，如果不想在某个地方显示，则设置为null。图标的宽和高自动设置为固定的宽和高，即自动通过getIntrinsicWidth和getIntrinsicHeight获取

 小贴士

在实际应用中，RadioButton 和 RadioGroup 通常配合使用。在没有 RadioGroup 的情况下，RadioButton 可以全部被选中；在多个 RadioButton 同时被 RadioGroup 包含的情况下，只可以选择一个 RadioButton，以达到单选的目的。RadioButton 和 RadioGroup 的关系体现为以下几点。

（1）RadioButton 表示单选按钮，而 RadioGroup 是可以容纳多个 RadioButton 的容器。

（2）每个 RadioGroup 中的多个 RadioButton 一次只能有一个被选中。

（3）不同的 RadioGroup 中的 RadioButton 互不相干，也就是说如果组 A 中有一个 RadioButton 选中了，那么组 B 中依然可以有一个 RadioButton 被选中。

（4）在大部分场合下，一个 RadioGroup 中至少有两个 RadioButton。

（5）在大部分场合下，一个 RadioGroup 中的 RadioButton 默认会有一个被选中，建议将这个被选中的 RadioButton 放在 RadioGroup 的起始位置。

4.2.4 Toast

Android 中的 Toast（消息框）用于向用户显示一些帮助信息或提示信息。Toast 一般通过下面两个步骤完成对消息框的显示。

（1）通过静态方法 makeText()创建一个 Toast 对象，即 Toast.makeText(context,text, duration)。其中，第一个参数是 Toast 要求的上下文；第二个参数是 Toast 显示的文本内容；第三个参数是 Toast 显示的时长，该参数有两个内置常量可供选择，即 Toast.LENGTH_SHORT 和 Toast.LENGTH_LONG。

（2）调用 show()方法显示 Toast，如 Toast.makeText(MainActivity.this,"hello",Toast.LENGTH_SHORT).show()。

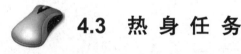

4.3 热身任务

本节热身任务为"寻找美女"。

1. 任务说明

（1）完成如图 4-2 所示的布局效果。

（2）实现功能：若用户选择的是"上海"，则显示"答对了"，如图 4-3 所示；若用户选择的是其他城市，则都显示"答错了"，如图 4-4 所示。

2. 操作步骤

（1）在 Android Studio 中新建一个项目，将其命名为 beauty。

（2）在 beauty.xml 文件中添加组件，产生如图 4-5 所示的效果。"寻找美女"的 Component Tree 如图 4-6 所示。

图4-2 "寻找美女"效果图

图4-3 选择正确答案后的效果图

图4-4 选择错误答案后的效果图

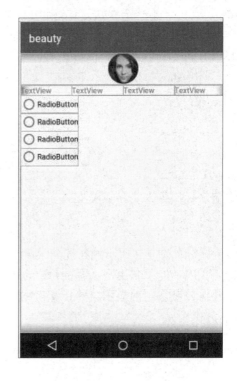

图4-5 初始布局

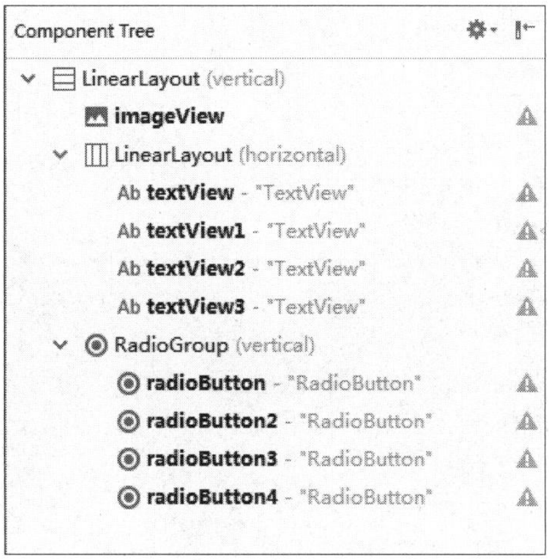

图 4-6 "寻找美女"的 Component Tree

（3）在项目的 res\drawable 文件夹中新建一个名为 back_shape.xml 的 shape 文件（创建过程可参照 3.2.7 节的相关内容），并添加相应属性，具体代码如下。

```
1.  <?xml version="1.0" encoding="utf-8"?>
2.  <shape xmlns:android="http://schemas.android.com/apk/res/android" >
3.      //通过添加以下属性可实现两种颜色的线性渐变效果
4.      <gradient android:startColor="#e86e5f" android:endColor="#f6e2c7"></gradient>
5.  </shape>
```

（4）将 ImageView 组件的 background 属性修改为第（3）步产生的 shape 文件（back_shape）。

（5）打开项目中的 res\values\styles.xml 文件，在该文件中添加一个样式 tv_face（添加该样式的原因是，4 个 TextView 组件相关组件属性的设置内容一致）。tv_face 样式的代码内容如下。

```
1.  <style name="tv_face">
2.      <item name="android:layout_width">0dp</item>
3.      <item name="android:textSize">16sp</item>
4.      <item name="android:layout_height">40dp</item>
5.      <item name="android:layout_weight">1</item>
6.      <item name="android:gravity">center</item>
7.      <item name="android:textColor">#fff</item>
8.  </style>
```

（6）将第（5）步创建的 tv_face 样式通过 TextView 组件的 style 属性分别应用到布局中的 4 个 TextView 组件。

（7）修改 beauty.xml 文件中相应组件的其他属性，修改完成后的 beauty.xml 文件的完整代码及相应功能说明如下（注意布局中修改了 4 个 RadioButton 的 ID）。

```
1.  <?xml version="1.0" encoding="utf-8"?>
2.  <LinearLayout xmlns:android="http://schemas.android.com/apk/res/android"
3.      xmlns:app="http://schemas.android.com/apk/res-auto"
4.      xmlns:tools="http://schemas.android.com/tools"
5.      android:layout_width="match_parent"
6.      android:layout_height="match_parent"
```

```
7.          android:orientation="vertical"
8.          tools:context=".MainActivity">
9.      <ImageView
10.         android:id="@+id/imageView"
11.         android:layout_width="match_parent"
12.         android:layout_height="59dp"
13.         android:background="@drawable/back_shape"
14.         app:srcCompat="@drawable/beauty" />
15.     <LinearLayout
16.         android:layout_width="match_parent"
17.         android:layout_height="wrap_content"
18.         android:gravity="center"
19.         android:layout_marginTop="5dp"
20.         android:layout_marginBottom="10dp">
21.         <TextView
22.             android:id="@+id/textView1"
23.             style="@style/tv_face"
24.             android:background="#f9ac98"
25.             android:text="25462 票" />
26.         <TextView
27.             android:id="@+id/textView2"
28.             style="@style/tv_face"
29.             android:background="#ff856d"
30.             android:text="54343 票" />
31.         <TextView
32.             android:id="@+id/textView3"
33.             style="@style/tv_face"
34.             android:background="#f9ac98"
35.             android:text="54432 票" />
36.         <TextView
37.             android:id="@+id/textView4"
38.             style="@style/tv_face"
39.             android:background="#ff856d"
40.             android:text="34545 票" />
41.     </LinearLayout>
42.     <TextView
43.         android:id="@+id/textView5"
44.         android:layout_width="wrap_content"
45.         android:layout_height="wrap_content"
46.         android:text="哪个城市美女多？" />
47.     <RadioGroup
48.         android:id="@+id/radioGroup1"
49.         android:layout_width="wrap_content"
50.         android:layout_height="wrap_content" >
51.         <RadioButton
52.             android:id="@+id/xAn"
53.             android:layout_width="wrap_content"
54.             android:layout_height="wrap_content"
55.             android:checked="true"//将 checked 设置为 true，让此单选按钮作为默认选中按钮
```

```
56.            android:text="西安" />
57.        <RadioButton
58.            android:id="@+id/sHai"
59.            android:layout_width="wrap_content"
60.            android:layout_height="wrap_content"
61.            android:text="上海" />
62.        <RadioButton
63.            android:id="@+id/gZhou"
64.            android:layout_width="wrap_content"
65.            android:layout_height="wrap_content"
66.            android:text="广州" />
67.        <RadioButton
68.            android:id="@+id/sZhen"
69.            android:layout_width="wrap_content"
70.            android:layout_height="wrap_content"
71.            android:text="深圳" />
72.    </RadioGroup>
73. </LinearLayout>
```

（8）打开 Java 文件夹下的 MainActivity.java 文件，在 onCreate()方法中为 RadioGroup1 组件添加 OnCheckedChangeListener 监听事件，并完成对选取答案正确与否的判断，具体代码如下。

```
1.    private RadioGroup rg;//声明 RadioGroup 组件的变量 rg
2.    protected void onCreate(Bundle savedInstanceState) {
3.        super.onCreate(savedInstanceState);
4.        setContentView(R.layout.activity_main);
5.        //将通过 ID 找到的单选按钮组赋值给 rg
6.        rg = (RadioGroup) this.findViewById(R.id.radioGroup1);
7.        //添加单选按钮组的选择改变监听
8.        rg.setOnCheckedChangeListener(new RadioGroup.OnCheckedChangeListener() {
9.            //i 保存用户选中的单选按钮的 ID。下面利用此变量进行比对，判断所选答案是否正确
10.           public void onCheckedChanged(RadioGroup radioGroup, int i) {
11.               switch (i) {
12.                   case R.id.xAn://如果 i 的值是 xAn，则说明用户选择了"西安"单选按钮
13.                       Toast.makeText(MainActivity.this, "答错了", Toast.LENGTH_SHORT).show();
14.                       break;
15.                   case R.id.gZhou:
16.                       Toast.makeText(MainActivity.this, "答错了", Toast.LENGTH_SHORT).show();
17.                       break;
18.                   case R.id.sHai:
19.                       Toast.makeText(MainActivity.this, "答对了", Toast.LENGTH_SHORT).show();
20.                       break;
21.                   case R.id.sZhen:
22.                       Toast.makeText(MainActivity.this, "答错了", Toast.LENGTH_SHORT).show();
23.                       break;
24.               }
25.           }
26.       });
27.   }
```

思考

如何为 RadioGroup 添加一个监听事件？
Toast. makeText()方法中第一个参数的上下文是什么？

 4.4 实现"底部导航"模块的布局

1. 知识点

➢ 线性布局的使用方法。
➢ RadioGroup 组件的使用方法。
➢ RadioButton 组件的使用方法。

2. 工作任务

制作"底部导航"模块的 UI 布局，其效果如图 4-7 所示。

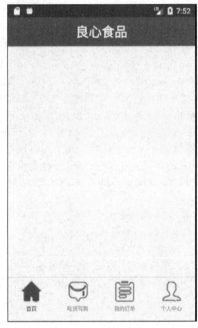

图 4-7　UI 布局效果

3. 操作流程

（1）依次单击"File"→"New"→"Activity"→"Empty Activity"选项，打开向导，完成"底部导航"页面的创建，创建过程使用默认文件名。

（2）在 activity_main.xml 文件中添加组件，产生如图 4-8 所示效果。导航页的 Component Tree 如图 4-9 所示。

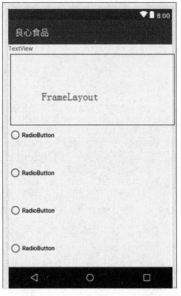

图 4-8　导航页初始布局

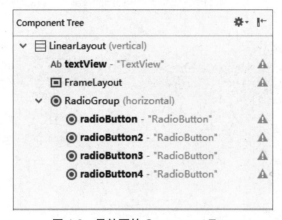

图 4-9　导航页的 Component Tree

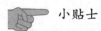 小贴士

在导航页初始布局中添加了一个 FrameLayout 布局用于显示主要的页面内容。FrameLayout 布局是最简单的布局，所有放在该布局里的控件，都按照层次堆叠在屏幕的左上角，后加进来的控件会覆盖前面的控件。

（3）在 Project 视图中打开项目的 res/values 文件夹下的 Dimens.xml 文件，并在该文件中添加 Navigation_RadioButtonsize 值，具体代码如下。

```
<dimen name="Navigation_RadioButtonsize">10sp</dimen>//定义底部导航单选按钮中文字的大小
```

（4）在 Project 视图中打开项目的 res/values 文件夹下的 colors.xml 文件，并在该文件中添加 public_green 等属性，具体代码如下。

```
1.    <color name='public_green'>#468500</color>
2.    <color name='navigation_false'>#60000000</color>
```

（5）在 Project 视图中打开项目的 res/values 文件夹下的 styles.xml 文件，并在该文件中添加 main_Radio Button 样式，用于设置底部导航单选按钮的外观样式，具体代码及其功能说明如下。

```
1.  <style name="main_RadioButton">
2.      <item name="android:layout_height">match_parent</item>//修改组件高度
3.      <item name="android:layout_width">0dp</item>//修改组件宽度
4.      <item name="android:layout_marginTop">4dp</item>
5.      <item name="android:layout_weight">1</item>//修改组件权重
6.      <item name="android:button">@null</item>//隐藏按钮样式
7.      <item name="android:drawablePadding">5dp</item>//修改图片内边距
8.      <item name="android:gravity">center_horizontal</item> //修改组件内部对齐方式
9.      <item name="android:textColor">@color/public_green</item>//修改组件中文字的颜色
10.     <item name="android:textSize">@dimen/Navigation_RadioButtonsize</item>//修改组件中文字的大小
11. </style>
```

（6）打开 activity_main.xml 文件，修改 TextView1 组件的属性，具体代码及其功能说明如下。

```
1.  <TextView
2.       android:id="@+id/main_title"//修改 id
3.       android:layout_width="match_parent"
4.       android:layout_height="0dp"
5.       android:layout_weight="1"
6.       android:background="@color/public_green"//修改组件背景
7.       android:gravity="center"
8.       android:text="@string/app_name"//修改组件中文字的内容
9.       android:textColor="#ffffff"
10.      android:textSize="25sp"
11.      android:textStyle="bold"//修改组件中文字的字体样式
12. />
```

（7）修改 FrameLayout 组件属性，具体代码如下。

```
1.  <FrameLayout
2.      android:id="@+id/main_framelayout"
3.      android:layout_width="match_parent"
4.      android:layout_height="0dp"
5.      android:layout_weight="10" >
6.  </FrameLayout>
```

（8）在 FrameLayout 组件与 RadioGroup 组件之间添加一个 View 并修改其属性，以实现添加一条直线的功能，具体代码如下。

```
1.  <View
2.      android:layout_width="match_parent"
3.      android:layout_height="1dp"
4.      android:background="#eceae9" />
```

（9）修改 RadioGroup 组件及 RadioButton 组件的属性，具体代码如下。

```
1.  <RadioGroup
2.      android:id="@+id/main_radioGroup"
```

```
3.          android:layout_width="match_parent"
4.          android:layout_height="0dp"
5.          android:layout_weight="1.5"
6.          android:gravity="center"
7.          android:orientation="horizontal">
8.          <RadioButton
9.              android:id="@+id/main_radio0"
10.             style="@style/main_RadioButton"
11.             android:checked="true"
12.             android:drawableTop="@mipmap/icon_home_true"
13.             android:text="首页"/>
14.         <RadioButton
15.             android:id="@+id/main_radio1"
16.             style="@style/main_RadioButton"
17.             android:drawableTop="@mipmap/icon_community_false"
18.             android:text="吃货驾到 "
19.             android:textColor="@color/navigation_false"/>
20.         <RadioButton
21.             android:id="@+id/main_radio2"
22.             style="@style/main_RadioButton"
23.             android:drawableTop="@mipmap/icon_order_false"
24.             android:textColor="@color/navigation_false"
25.             android:text="我的订单"/>
26.         <RadioButton
27.             android:id="@+id/main_radio3"
28.             style="@style/main_RadioButton"
29.             android:drawableTop="@mipmap/icon_me_false"
30.             android:textColor="@color/navigation_false"
31.             android:text="个人中心"/>
32.     </RadioGroup>
```

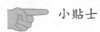 小贴士

按照以上操作流程制作的布局还会多一个标题栏，这时可将样式文件中的 AppTheme 的 parent 修改为 Theme.AppCompat.Light.NoActionBar。

 4.5 实现导航功能

1. 知识点

- ➢ RadioGroup 的 OnCheckedChangeListener 监听事件的使用方法。
- ➢ 动态获取 drawable 文件夹下图片资源的方法。
- ➢ 动态修改 RadioButton 中文字字体的方法。
- ➢ 动态修改 RadioButton 内置图片的方法。
- ➢ Toast 的使用方法。

2. 工作任务

当单击底部导航选项按钮时，被选中的选项图片及文字颜色改变的同时，弹出消息框显示导航标题，如图 4-10 和图 4-11 所示。

图 4-10 "首页"选项被选中的效果图

图 4-11 "吃货驾到"选项被选中的效果图

3. 操作流程

（1）在 Project 视图中打开项目的 java 文件夹下的 MainActvity.java 源程序文件。

（2）在源程序中定义相关变量，具体代码及其功能说明如下。

```
1.   private RadioGroup myradioGroup;         //单选按钮组
2.   private RadioButton rbutton1, rbutton2, rbutton3, rbutton4; //4 个单选按钮
3.   private Resources res;
4.   private Drawable icon_home_true, icon_home_false, icon_me_true,
5.       icon_me_false, icon_order_true, icon_order_false,
6.       icon_community_true, icon_community_false;   //4 个单选按钮选中和未选中时的图标
7.   private int fontColor_false, fontColor_true; //4 个单选按钮选中和未选中时的文字颜色
```

（3）在源程序中添加 initview()方法用于初始化 UI 组件，以及进行图片、颜色等的相关准备工作，具体代码及其功能说明如下。

```
1.   private void initview() {
2.       //通过 ID 找到 UI 中的单选按钮组
3.       myradioGroup = (RadioGroup) this.findViewById(R.id.main_radioGroup);
4.       rbutton1 = (RadioButton) this.findViewById(R.id.main_radio0); //通过 ID 找到 UI 中的单选按钮
5.       rbutton2 = (RadioButton) this.findViewById(R.id.main_radio1);
6.       rbutton3 = (RadioButton) this.findViewById(R.id.main_radio2);
7.       rbutton4 = (RadioButton) this.findViewById(R.id.main_radio3);
8.       res = getResources();//得到 Resources 对象，通过它获取存于系统的资源
9.       //找到图片 icon_home_true，用于设置当"首页"选项被选中时的图片
10.      icon_home_true = res.getDrawable(R.mipmap.icon_home_true);
11.      //找到图片 icon_home_false，用于设置当"首页"选项未被选中时的图片
12.      icon_home_false = res.getDrawable(R.mipmap.icon_home_false);
13.          //找到图片 icon_community_true，用于设置当"吃货驾到"选项被选中时的图片
```

```
14.         icon_community_true = res.getDrawable(R.mipmap.icon_community_true);
15.         //找到图片 icon_community_false,用于设置当"吃货驾到"选项未被选中时的图片
16.         icon_community_false = res.getDrawable(R.mipmap.icon_community_false);
17.         //找到图片 icon_me_true,用于设置当"个人中心"选项被选中时的图片
18.         icon_me_true = res.getDrawable(R.mipmap.icon_me_true);
19.         //找到图片 icon_me_false,用于设置当"个人中心"选项未被选中时的图片
20.         icon_me_false = res.getDrawable(R.mipmap.icon_me_false);
21.         //找到图片 icon_order_true,用于设置当"我的订单"选项被选中时的图片
22.         icon_order_true = res.getDrawable(R.mipmap.icon_order_true);
23.         //找到图片 icon_order_false,用于设置当"我的订单"选项未被选中时的图片
24.         icon_order_false = res.getDrawable(R.mipmap.icon_order_false);
25.         //找到颜色 navigation_false,用于设置当选项未被选中时的文字颜色
26.         fontColor_false = res.getColor(R.color.navigation_false);
27.         //找到颜色 public_green,用于设置当选项被选中时的文字颜色
28.         fontColor_true = res.getColor(R.color.public_green);
29.     }
```

（4）在源程序中添加 setAllColor()方法，用于将所有选项的文字颜色设置为未被选中状态下的文字颜色 fontColor_false，具体代码如下。

```
1.  private void setAllColor(){
2.      rbutton1.setTextColor(fontColor_false);
3.      rbutton2.setTextColor(fontColor_false);
4.      rbutton3.setTextColor(fontColor_false);
5.      rbutton4.setTextColor(fontColor_false);
6.  }
```

（5）在源程序中添加 setAllImage()方法，用于将所有选项的图片设置为未被选中状态下的图片，具体代码及其功能说明如下。

```
1.  private void setAllImage(){
2.      //设置"首页"选项图片为未被选中时的图片
3.      rbutton1.setCompoundDrawablesWithIntrinsicBounds(null, icon_home_false, null, null);
4.      //设置"吃货驾到"选项图片为未被选中时的图片
5.      rbutton2.setCompoundDrawablesWithIntrinsicBounds(null, icon_community_false, null, null);
6.      //设置"我的订单"选项图片为未被选中时的图片
7.      rbutton3.setCompoundDrawablesWithIntrinsicBounds(null, icon_order_false, null, null);
8.      //设置"个人中心"选项图片为未被选中时的图片
9.      rbutton4.setCompoundDrawablesWithIntrinsicBounds(null, icon_me_false, null, null);
10. }
```

（6）在源程序中添加 navigation()方法，用于实现"底部导航"选项选中状态与未选中状态切换的功能，具体代码及其功能说明如下。

```
1.  private void navigation() {
2.      // TODO Auto-generated method stub
3.      myradioGroup.setOnCheckedChangeListener(new RadioGroup.OnCheckedChangeListener(){
4.          //变量 int i 中保存了用户每次选中的选项的 ID,下面的操作就是利用此特点来确定单选按钮被选中的状态,并实现相应的需求的
5.          public void onCheckedChanged(RadioGroup radioGroup, int i) {
6.              //调用此方法用于在每次切换选项时将所有选项的文字颜色复位为未被选时的字体颜色
7.              setAllColor();
```

```
8.              //调用此方法用于在每次切换选项时将所有选项的图片复位为未被选时的图片
9.              setAllImage();
10.             switch (i) {
11.                 //当"首页"选项被选中时,设置选项在选中状态下的文字及图片
12.                 case R.id.main_radio0:
13.                     rbutton1.setTextColor(fontColor_true);
14.                     rbutton1.setCompoundDrawablesWithIntrinsicBounds(null, icon_home_true, null, null);
15.                     Toast.makeText(MainActivity.this,"首页",Toast.LENGTH_SHORT). show();
16.                     break;
17.                 //当"吃货驾到"选项被选中时,设置选项在选中状态下的文字及图片
18.                 case R.id.main_radio1:
19.                     rbutton2.setTextColor(fontColor_true);
20.                     rbutton2.setCompoundDrawablesWithIntrinsicBounds(null, icon_community_true, null, null);
21.                     Toast.makeText(MainActivity.this,"吃货驾到",Toast.LENGTH_SHORT). show();
22.                     break;
23.                 //当"我的订单"选项被选中时,设置选项在选中状态下的文字及图片
24.                 case R.id.main_radio2:
25.                     rbutton3.setTextColor(fontColor_true);
26.                     rbutton3.setCompoundDrawablesWithIntrinsicBounds(null, icon_order_true, null, null);
27.                     Toast.makeText(MainActivity.this,"我的订单",Toast.LENGTH_SHORT).show();
28.                     break;
29.                 //当"个人中心"选项被选中时,设置选项在选中状态下的文字及图片
30.                 case R.id.main_radio3:
31.                     rbutton4.setTextColor(fontColor_true);
32.                     rbutton4.setCompoundDrawablesWithIntrinsicBounds(null, icon_me_true, null, null);
33.                     Toast.makeText(MainActivity.this,"个人中心", Toast.LENGTH_SHORT).show();
34.                     break;
35.             }
36.         }
37.     });
38. }
```

(7)在源程序中重写 onCreate()方法以调用初始化方法 initview()及实现底部导航功能的方法 navigation(),具体代码及其功能说明如下。

```
1. protected void onCreate(Bundle savedInstanceState) {
2.     super.onCreate(savedInstanceState);
3.     setContentView(R.layout.activity_main);
4.     initview();//调用此方法实现初始化组件的功能
5.     navigation();//调用此方法实现设置单选按钮组选项改变监听事件的功能从而实现导航功能
6. }
```

(8)运行程序之前需要将 AndroidManifest.xml 中的<intent-filter>与</intent-filter>之间的代码由原来的 name 值为 Login 的 Activity 移至 name 值为 MainActivity 的 Activity 中,从而使应用启动页面由原来的"登录"页面改为"底部导航"页面。修改后的 AndroidManifest.xml 代码如下。

```
1. <?xml version="1.0" encoding="utf-8"?>
```

```
2.   <manifest xmlns:android="http://schemas.android.com/apk/res/android"
3.       package="com.example.myfrist.healthfood">
4.       <application
5.           android:allowBackup="true"
6.           android:icon="@mipmap/ic_launcher"
7.           android:label="@string/app_name"
8.           android:roundIcon="@mipmap/ic_launcher_round"
9.           android:supportsRtl="true"
10.          android:theme="@style/AppTheme">
11.          <activity android:name=".Login">
12.          </activity>
13.          <activity android:name=".MainActivity">
14.              <intent-filter>
15.                  <action android:name="android.intent.action.MAIN" />
16.                  <category android:name="android.intent.category.LAUNCHER" />
17.              </intent-filter>
18.          </activity>
19.      </application>
20.  </manifest>
```

（9）运行程序。

 小贴士

在进行"导航"选项选中状态和未选中状态切换时，主要采用了当触发了单选按钮组的 onCheckedChanged 监听事件后，先将所有的单选按钮恢复成未选中状态（用 setAllImage()方法来实现将所有的单选按钮的内置图标恢复成未被选中时的图标，用 setAllColor()方法来实现将所有的单选按钮的文字颜色恢复成未被选中时的文字颜色），然后通过判断将被选中的单选按钮修改为选中状态（单选按钮中的图片和文字的颜色都变为绿色）的方法。

第 5 章 "个人中心"模块的设计

 教学目标

✧ 掌握 Fragment 的生命周期。
✧ 掌握创建 Fragment 的方法。
✧ 掌握动态加载 Fragment 的方法。

 ## 5.1 工作任务概述

App 的"个人中心"是私人化的用户界面,用户可以在该界面查看"个人信息",并可以进行相关自定义的设置。本章工作任务是完成"个人中心"的制作,需要完成以下工作子任务。
(1)完成"个人中心"的 UI 布局,效果如图 5-1 所示。
(2)制作"个人中心"的 Fragment。
(3)将"个人中心"添加到有底部导航的主 Activity 框架内(第 4 章创建的 MainActivity)。

图 5-1 "个人中心"效果图

5.2 预备知识

5.2.1 Fragment

1. Fragment 简介

Fragment 是一个 Android3.0 版本之后引入的 API，主要用于为大屏幕的平板电脑实现动态、灵活的界面设计提供支持，普通手机的开发也可以加入 Fragment。Fragment 又称 Activity 片段，可以将 Fragment 理解为小型的 Activity。如果一个很大的界面只有一个布局，在设计界面时会非常麻烦，若组件有很多，管理起来也很麻烦。而使用 Fragment，可以把屏幕划分成几块，然后分组，进行模块化的管理，这样可以更加方便地在运行过程中动态更新 Activity 的用户界面。Fragment 并不能单独使用，而需要嵌套在 Activity 中使用。尽管 Fragment 拥有自己的生命周期，但其还是会受到宿主 Activity 的生命周期的影响，比如，若 Activity 被销毁，Fragment 也会被销毁。

2. Fragment 的生命周期图（见图 5-2）

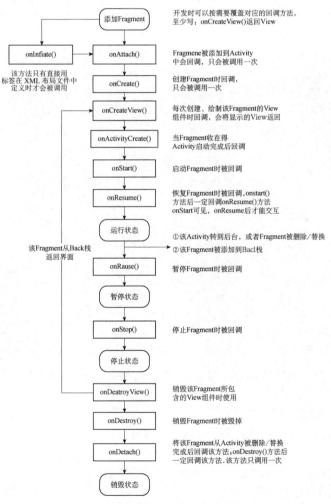

图 5-2　Fragment 的生命周期图

3. FragmentTransaction 常用方法及其作用（见表 5-1）

表 5-1　FragmentTransaction 常用方法及其作用

方　　法	作　　用
replace()	使用另一个Fragment替换当前的Fragment
add(Fragment,String)	添加一个没有UI的Fragment，该Fragment只能通过FragmentManager.findFragmentByTag()获取
add(int,Fragment)	添加一个Fragment，参数1用于添加Fragment的ID，参数2用于添加Fragment实例
remove()	从Activity中移除一个Fragment
hide()	隐藏当前的Fragment，只是将该Fragment设置为不可见，并不会将其销毁
show()	显示之前隐藏的Fragment
detach()	将此Fragment从Activity中分离，会销毁其布局，但不会销毁该实例
attach()	将从Activity中分离的Fragment重新关联到该Activity，重新创建其视图层次
addToBackStack()	添加事务到回退栈
commit()	提交一个事务。一个事务从开启到提交可以进行多次添加、移除、替换等操作

4. 创建 Fragment 的两种方式

（1）在 XML 布局中使用标签。

```
1.  <?xml version="1.0" encoding="utf-8"?>
2.  <LinearLayout xmlns:android="http://schemas.android.com/apk/res/android"
3.      android:orientation="horizontal"
4.      android:layout_width="match_parent"
5.      android:layout_height="match_parent">
6.      <fragment android:name="com.example.news.ArticleListFragment"
7.          android:id="@+id/list"
8.          android:layout_weight="1"
9.          android:layout_width="0dp"
10.         android:layout_height="match_parent" />
11.     <fragment android:name="com.example.news.ArticleReaderFragment"
12.         android:id="@+id/viewer"
13.         android:layout_weight="2"
14.         android:layout_width="0dp"
15.         android:layout_height="match_parent" />
16. </LinearLayout>
```

用此方式创建的 Fragment 不能被移除，而且 Fragment 与 Activity 同时被创建，灵活性差。Fragment 仅作为简单的视图展示时可以使用该方式。

（2）代码动态添加。

```
1.  FragmentManager manager = getFragmentManager()
2.  FragmentTransaction transaction = manager.beginTransaction();
3.  ExampleFragment fragment = new ExampleFragment();
4.  transaction.add(R.id.fragment_container, fragment);
5.  transaction.commit();
```

用此方式创建的 Fragment 可以实现不同 Fragment 之间的切换,可以实现 Activity 运行时对 Fragment 进行添加、移除和替换,可控性高,方便处理不同屏幕的适配,但布局中必须有一个视图容器用来存放 Fragment。

5. 动态创建一个 Fragment 的步骤

(1)定义 Fragment 的 XML 布局文件。
(2)自定义 Fragment 类,需要继承 Fragment 类或其子类,同时重写 onCreateView()方法。
(3)通过 getFragmentManager()方法获得 FragmentManager 对象。
(4)通过 FragmentManage.beginTransaction()方法获得 FragmentTransaction 对象。
(5)调用 add()方法或 replace()方法加载 Fragment。
(6)在前面步骤的基础上调用 commit()方法(或 remove()方法等)提交事务。

 小贴士

Transaction.replace(containerViewId,fragment)方法中的第一个参数 containerViewId 表示 Fragment 要放入的 ViewGroup 的资源 ID,第二个参数 fragment 表示要替换的 Fragment。

5.2.2 Intent

1. Intent 概述

在 Android 应用软件的开发过程中,Intent 是一个非常重要的类,用来启动或加载 Android 应用程序组件,并且通过它能够在不同的组件之间传输数据。Intent 主要有以下几种重要用途。

(1)启动 Activity:可以将 Intent 对象传递给 startActivity()方法或 startActivityForResult()方法以启动一个 Activity,该 Intent 对象包含要启动的 Activity 的信息及其他必要的数据。

(2)启动 Service:可以将 Intent 对象传递给 startService()方法或 bindService()方法以启动一个 Service,该 Intent 对象包含要启动的 Service 的信息及其他必要的数据。

(3)发送广播:广播是一种所有 App 都可以接收的信息。Android 会发布各种类型的广播,如开机广播或手机充电广播等;也可以给其他 App 发送广播;还可以将 Intent 对象传递给 sendBroadcast()方法、sendOrderedBroadcast()方法或 sendStickyBroadcast()方法以发送自定义广播。

在 Android 应用程序中,Intent 通过 Action、Category 和 Data 等属性对 Android 应用程序进行相应描述,若需要实现某些功能,则要填写这些属性的部分内容或全部内容,这样 Android 应用程序才会自动去进行某些操作。IntentFilter 类专门用来配合 Intent 过滤要启动的组件。

2. Intent 的类型

Intent 有两种类型,即显式(Explicit)Intent 和隐式(Implicit)Intent。

(1)显式 Intent。如果 Intent 中明确包含要启动的组件的完整类名(包名及类名),那么这个 Intent 就是显式的。例如,若要启动一个特定的 Activity,只需要将当前的 Context 和该 Activity 的 class 作为参数构造 Intent,然后再将这个 Intent 作为参数传递给 startActivity()方法即可,具体代码如下。

1. Intent intent=new Intent(MainActivity.this,OtherActivity.class);

```
2.    startActivity(intent)
```

（2）隐式 Intent。如果 Intent 不包含要启动的组件的完整类名，那么这个 Intent 就是隐式的。例如，对于第三方的 Activity，它描述自己在什么情况下被启动，如果启动 Activity 的描述信息正好和第三方 Activity 的描述相匹配，那么这个第三方的 Activity 就会被启动，具体代码如下。

```
1.    Intent intent=new Intent(Intent.ACTION_CALL,"tel:18000100011");
2.    startActivity(intent)
```

 小贴士

Intent 显式跳转需要指定具体要跳转的目标。本教程中的 Activity 之间的跳转使用显式跳转。Intent 隐式跳转可以在自己的应用程序中启动其他程序的 Activity，这样可以使多个应用程序的功能实现共享。通过 Intent 隐式跳转可以调用拨号面板、发送短信息等功能模块。

3. 多个 Activity 和 Intent

Android 应用程序在启动时，系统首先执行 Androidmanifest.xml 文件中定义的主 Activity，然后在主 Activity 中通过事件（如按钮事件）触发启动并跳转到另一个 Activity。当 Activity 被激活并运行时，它将处于 Activity 栈的顶部，而之前处于 Activity 栈顶部的 Activity 将被压入栈中处于暂停状态；当新的 Activity 运行结束后，便将控制权交回先前的 Activity，此时之前的 Activity 将重新回到前台恢复运行。按照这样的方式不断重复，可以为 Android 应用程序创建若干个 Activity 并在 Activity 之间进行两两跳转，而 Intent 在这些 Activity 之间充当桥梁，为它们之间的跳转传递各种消息。

通常，启动另一个 Activity 及在 Activity 之间跳转的步骤如下。

第一步：定义一个 Intent，并为该 Intent 指定即将被启动的 Activity。

第二步：调用 Intent 的 startActivity()方法，启动并跳转到新的 Activity。

例如，有两个 Activity，分别为 Activity1 和 Activity2，需要从 Activity1 跳转到 Activity2，具体实现代码如下。

```
1.    Intent intent = new Intent(Activity1.this, Activity2.class);
2.    startActivity(intent);
```

在上述代码中，在未调用 startActivity()方法时，Activity1 作为当前 Activity 会处于 Activity 栈的顶部，如图 5-3（a）所示；当调用 startActivity()方法后，Activity2 处于 Activity 栈的顶部，如图 5-3（b）所示；当 Activity2 结束时，在后台的 Activity1 又会被提到前台来执行，如图 5-3（c）所示。startActivity(Intent)方法可以启动新的 Activity，而 finish()方法则可以结束当前的 Activity。

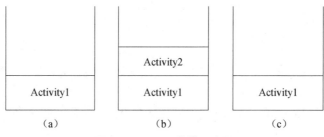

图 5-3　Activity 堆栈示意图

5.3 热身任务

本节热身任务为"王者技能"。

1. 任务说明

要求通过 Fragment 完成以下功能：当单击"技能简介"下的第一个图片时，显示"咒术火焰"的技能简介，如图 5-4 所示；当单击第二个图片时，则显示"火球术"的技能简介，如图 5-5 所示。

图5-4 "咒术火焰"的技能简介图

图5-5 "火球术"的技能简介图

2. 操作步骤

（1）创建一个 Android 项目，项目名自定。

（2）制作如图 5-6 所示的布局，图 5-6 中指向"FrameLayout"的箭头部分使用的是 FrameLayout 布局。"王者技能"的 Component Tree 如图 5-7 所示。

图5-6 "王者技能"布局图

图5-7 "王者技能"的 Component Tree

"王者技能"的 xml 代码如下。

```
1.  <android.support.constraint.ConstraintLayout
        xmlns:android="http://schemas.android.com/apk/res/android"
2.      xmlns:app="http://schemas.android.com/apk/res-auto"
3.      xmlns:tools="http://schemas.android.com/tools"
4.      android:layout_width="match_parent"
5.      android:layout_height="match_parent"
6.      tools:context=".MainActivity">
7.      <ImageView
8.          android:id="@+id/imageView"
9.          android:layout_width="wrap_content"
10.         android:layout_height="214dp"
11.         app:layout_constraintStart_toStartOf="parent"
12.         app:layout_constraintTop_toTopOf="parent"
13.         app:srcCompat="@drawable/chapter4a" />
14.     <TextView
15.         android:id="@+id/textView"
16.         android:layout_width="wrap_content"
17.         android:layout_height="wrap_content"
18.         android:layout_marginStart="8dp"
19.         android:layout_marginTop="10dp"
20.         android:text="技能简介"
21.         android:textColor="#000"
22.         android:textSize="20sp"
23.         app:layout_constraintStart_toStartOf="parent"
24.         app:layout_constraintTop_toBottomOf="@id/imageView" />
25.     <ImageView
26.         android:id="@+id/imageView2"
27.         android:layout_width="50dp"
28.         android:layout_height="50dp"
29.         android:layout_marginTop="10dp"
30.         app:layout_constraintTop_toBottomOf="@id/textView"
31.         app:layout_constraintStart_toStartOf="@id/textView"
32.         app:srcCompat="@drawable/p1"/>
33.     <ImageView
34.         android:id="@+id/imageView3"
35.         android:layout_width="50dp"
36.         android:layout_height="50dp"
37.         app:srcCompat="@drawable/p2"
38.         android:layout_marginLeft="10dp"
39.         app:layout_constraintTop_toTopOf="@id/imageView2"
40.         app:layout_constraintStart_toEndOf="@id/imageView2"/>
41.     <FrameLayout
42.         android:layout_width="match_parent"
43.         android:layout_height="201dp"
44.         android:layout_marginTop="10dp"
45.         app:layout_constraintStart_toStartOf="parent"
46.         app:layout_constraintTop_toBottomOf="@id/imageView2">
```

```
47.        </FrameLayout>
58.    </android.support.constraint.ConstraintLayout>
```

（3）将光标移至 layout 文件夹上并右击，依次单击"New"→"XML"→"Layout XML File"选项，新建一个名为 hqs.xml 的 XML 布局文件，制作如图 5-8 所示的布局，XML 布局文件代码如下。

```
1.  <?xml version="1.0" encoding="utf-8"?>
2.  <LinearLayout xmlns:android="http://schemas.android.com/apk/res/android"
3.      android:layout_width="match_parent"
4.      android:layout_height="match_parent"
5.      android:orientation="vertical" >
6.      <TextView
7.          android:id="@+id/textView1"
8.          android:layout_width="wrap_content"
9.          android:layout_height="wrap_content"
10.         android:textSize="20sp"
11.         android:text="火球术" />
12.     <TextView
13.         android:id="@+id/textView2"
14.         android:layout_width="wrap_content"
15.         android:layout_height="wrap_content"
16.         android:text="安琪拉召唤 5 颗火球朝指定位置攻击，每颗火球对敌人造成 350/390/430/470/510/550（+30%法术加成）点法术伤害。当敌人被多颗火球命中时，额外的火球只能造成 30%伤害。火球命中英雄时会停止移动并销毁。" />
17. </LinearLayout>
```

图 5-8 "火球术"的技能简介碎片布局图

（4）在项目的 java 文件夹的源程序文件夹中添加一个 Fragment_hqs.java 类，此类继承 Fragment。

（5）完成 Fragment_hqs.java 类的创建后，该类中无任何代码，此时可通过右击（或快捷键 Alt+Insert）并依次单击"Generate"→"Override Method"选项，将 onCreateView()方法添加到 java 源程序中，同时重写 Fragment_hqs.java 类中的 onCreateView()方法，利用 onCreateView()方法中的布局加载器 inflater 将第（3）步创建的"火球术"技能简介碎片布局组装到碎片中，具体代码如下。

```
1.  public class Fragment_hqs extends Fragment {
2.      public View onCreateView(LayoutInflater inflater, ViewGroup container, Bundle savedInstanceState) {
3.          return inflater.inflate(R.layout.hqs, container, false);
```

```
4.        //R.layout.hqs 为碎片的 XML 布局文件
5.        //container 存放 Fragment 的 Layout 的 ViewGroup
6.        //布尔数据表示是否在创建 Fragment 的 layout 文件夹期间，把 layout 文件夹附加到
container 上（在这个例子中，因为系统已经把 layout 文件夹插入 container 中了，所以布尔数据
的值为 false，如果值为 true 会导致在最终的 layout 文件夹中创建多余的 ViewGroup）
7.
8.     }
9. }
```

（6）将光标移至 layout 文件夹上并右击，依次单击"New"→"XML"→"Layout XML File"选项，新建一个名为 zshy.xml 的 XML 布局文件，制作如图 5-9 所示的布局，布局文件如下。

```
1. <?xml version="1.0" encoding="utf-8"?>
2. <LinearLayout xmlns:android="http://schemas.android.com/apk/res/android"
3.     android:layout_width="match_parent"
4.     android:layout_height="match_parent"
5.     android:orientation="vertical" >
6.     <TextView
7.         android:id="@+id/textView1"
8.         android:layout_width="wrap_content"
9.         android:layout_height="wrap_content"
10.        android:textSize="20sp"
11.        android:text="咒术火焰" />
12.    <TextView
13.        android:id="@+id/textView2"
14.        android:layout_width="wrap_content"
15.        android:layout_height="wrap_content"
16.        android:text="咒术火焰是安琪拉的核心技能，她能够使安琪拉的所有技能伤害都附加一个额外魔法伤害，依靠这个被动技能效果，安琪拉在施放技能持续攻击敌人时，能够造成成吨的伤害。" />
17. </LinearLayout>
```

图 5-9 "咒术火焰"的技能简介碎片布局图

（7）创建一个 Fragment_zshy.java 类（操作方法与第（5）步相同），此类继承 Fragment，同时重写该类中的 onCreateView()方法，利用 onCreateView()方法中的布局加载器 inflater 将第（6）步创建的"咒术火焰"技能简介碎片布局组装到碎片中，具体代码如下。

```
1. public class Fragment_zshy extends Fragment {
2.     @Override
3.     public View onCreateView(LayoutInflater inflater, ViewGroup container,
Bundle savedInstanceState) {
4.         // TODO Auto-generated method stub
```

```
5.            return inflater.inflate(R.layout.zshy, container, false);
6.        }
7.  }
```

(8) 在项目的 java 文件夹中找到 MainActivity.java 文件，打开该文件并在其中添加相应的代码完成碎片的切换。

在程序中创建 FragmentChange()方法，用于加载和替换 Fragment 碎片，具体代码如下。

```
1.  public void FragmentChange(Fragment myFragment) {
2.      FragmentManager manager = this.getFragmentManager();//获取 FragmentManager
3.      FragmentTransaction transaction = manager.beginTransaction();//开启一个碎片管理事务
4.      // R.id.fragment_container 是主布局的 FrameLayout 组件，主要用于显示碎片；myFragment
        是要显示的碎片
5.      transaction.replace(R.id.fragment_container, myFragment);
6.      transaction.commit();//提交事务请求
7.  }
```

(9) 在 MainActivity.java 文件中添加 init()方法，该方法用于初始化组件并为两张用于切换的图片添加单击监听，以实现加载相应碎片，具体代码如下（第 8~13 行代码用于对碎片加载与替换）。

```
1.  public class MainActivity extends Activity {
2.      ImageView Img_hqs, Img_zshy;
3.      protected void onCreate(Bundle savedInstanceState) {
4.          super.onCreate(savedInstanceState);
5.          setContentView(R.layout.activity_main);
6.          init();
7.      }
8.      public void FragmentChange(Fragment myFragment) {
9.          FragmentManager manager = this.getFragmentManager();//获取 FragmentManager
10.         FragmentTransaction transaction = manager.beginTransaction();//开启一个碎片管理事务
11.         //R.id.fragment_container 是主布局的 FrameLayout 组件，主要用于显示碎片；myFragment
            是要显示的碎片
12.         transaction.replace(R.id.fragment_container,myFragment);
13.         transaction.commit();//提交事务请求
14.     }
15.     public void init() {
16.         //在布局文件中找到用于显示"火球术"技能的开关图片组件
17.         Img_hqs = (ImageView) this.findViewById(R.id.imageView2);
18.         //在布局文件中找到用于显示"咒术火焰"技能的开关图片组件
19.         Img_zshy = (ImageView) this.findViewById(R.id.imageView3);
20.         //为"火球术"技能的开关图片添加单击监听事件，用于显示相对应的碎片
21.         Img_hqs.setOnClickListener(new View.OnClickListener() {
22.             @Override
23.             public void onClick(View view) {
24.                 Fragment_hqs myFragment = new Fragment_hqs();//创建"火球术"技能简介碎片
25.                 //调用碎片加载方法，将"火球术"技能简介界面显示于页面
26.                 FragmentChange(myFragment);
27.             }
28.         });
29.         //为"咒术火焰"技能的开关图片添加单击监听事件，用于显示相对应的碎片
```

```
30.        Img_zshy.setOnClickListener(new View.OnClickListener() {
31.            @Override
32.            public void onClick(View view) {
33.                Fragment_zshy myFragment = new Fragment_zshy();//创建"咒术火焰"技能简介碎片
34.                //调用碎片加载方法,将"咒术火焰"技能简介界面在页面中进行显示
35.                FragmentChange(myFragment);
36.            }
37.        });
38.    }
39. }
```

 思考

本案例中的布局为什么要使用 FrameLayout 组件?

 5.4　实现"个人中心"模块的布局

1. 知识点

➢ 线性布局。
➢ ConstraintLayout 布局。
➢ 样式文件 style.xml。

2. 工作任务

制作"个人中心"模块的 UI 布局,产生如图 5-10 所示的 UI 效果。

图 5-10　"个人中心" UI 效果图

3. 操作流程

（1）依次单击"File"→"New"→"XML"→"Layout XML File"选项，创建 frag_personalcenter.xml 文件。

（2）在 frag_personalcenter.xml 文件中添加组件，产生如图 5-11 所示效果。"个人中心"的 Component Tree 如图 5-12 所示。

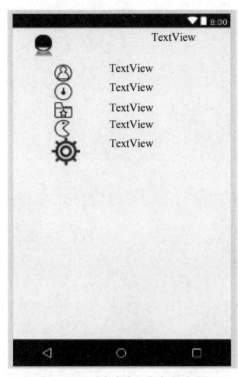

图 5-11　"个人中心"初始布局

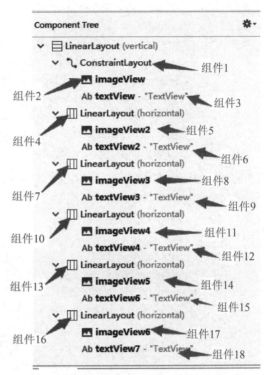

图 5-12　"个人中心"的 Component Tree

（3）打开项目 res/values 文件夹下的 style.xml 文件，并在该文件中添加 personal_Liner 样式，具体代码及其功能说明如下。

```
1.    <style name="personal_Liner">
2.        <item name="android:background">#fff</item>//设置背景颜色
3.        <item name="android:layout_marginBottom">2dp</item>//设置底部与下边组件的距离
4.        <item name="android:padding">5dp</item>//设置内边距
5.        <item name="android:gravity">center_vertical</item>//设置组件内元素的对齐方式
6.    </style>
```

（4）打开 frag_personalcenter.xml 文件，修改图 5-12 中组件 1 的属性，具体代码如下。

```
1.    <android.support.constraint.ConstraintLayout
2.        android:layout_width="match_parent"
3.        android:layout_height="60dp"
4.        android:layout_marginTop="10dp"
5.        android:layout_marginBottom="10dp"
6.        android:background="#fff" >
```

(5) 修改图 5-12 中组件 2 的属性，具体代码如下。

```
1.  <ImageView
2.      android:id="@+id/imageView"
3.      android:layout_width="50dp"
4.      android:layout_height="50dp"
5.      android:layout_marginBottom="8dp"
6.      android:layout_marginEnd="8dp"
7.      android:layout_marginStart="8dp"
8.      android:layout_marginTop="8dp"
9.      app:layout_constraintBottom_toBottomOf="parent"
10.     app:layout_constraintEnd_toEndOf="parent"
11.     app:layout_constraintHorizontal_bias="0.025"
12.     app:layout_constraintStart_toStartOf="parent"
13.     app:layout_constraintTop_toTopOf="parent"
14.     app:srcCompat="@mipmap/login_userpic1" />
```

(6) 修改图 5-12 中组件 3 的属性，具体代码如下。

```
1.  <TextView
2.      android:id="@+id/textView"
3.      android:layout_width="wrap_content"
4.      android:layout_height="wrap_content"
5.      android:layout_marginEnd="8dp"
6.      android:text="待君登录"
7.      app:layout_constraintBottom_toBottomOf="parent"
8.      app:layout_constraintEnd_toEndOf="parent"
9.      app:layout_constraintHorizontal_bias="0.209"
10.     app:layout_constraintStart_toStartOf="@+id/imageView"
11.     app:layout_constraintTop_toTopOf="parent"
12.     app:layout_constraintVertical_bias="0.512" />
```

(7) 修改图 5-12 中组件 4 的属性，具体代码如下。

```
1.  <LinearLayout
2.      android:layout_width="match_parent"
3.      android:layout_height="wrap_content"
4.      style="@style/personal_Liner"
5.      android:orientation="horizontal">
```

(8) 修改图 5-12 中组件 5 的属性，具体代码如下。

```
1.  <ImageView
2.      android:id="@+id/imageView2"
3.      android:layout_width="wrap_content"
4.      android:layout_height="wrap_content"
5.      android:layout_weight="1"
6.      app:srcCompat="@mipmap/icmember1" />
```

(9)修改图 5-12 中组件 6 的属性,具体代码如下。

```
1.    <TextView
2.        android:id="@+id/textView2"
3.        android:layout_width="wrap_content"
4.        android:layout_height="wrap_content"
5.        android:layout_weight="4"
6.        android:text="个人信息" />
```

(10)对图 5-12 中的组件 7、组件 10、组件 13 分别进行与组件 4 相同的属性修改。

(11)对图 5-12 中的组件 8、组件 11、组件 14、组件 17 分别进行与组件 5 相类似的属性修改,唯一不同的是将 app:srcCompat 修改为个性图片源。

(12)对图 5-12 中的组件 9、组件 12、组件 15、组件 18 分别进行与组件 6 相类似的属性修改,唯一不同的是将 android:text 修改成个性标题。

(13)修改图 5-12 中的组件 16 的属性,具体代码如下。

```
1.    <LinearLayout
2.        android:layout_width="match_parent"
3.        android:layout_height="wrap_content"
4.        style="@style/personal_Liner"
5.        android:layout_marginTop="20dp"
6.        android:orientation="horizontal">
```

5.5 创建"个人中心"Fragment

1. 知识点

Fragment 的创建方法。

2. 工作任务

在 5.4 节工作任务的基础上将 frag_personalcenter.xml 文件转换成"个人中心"碎片。

3. 操作流程

(1)右击项目中 java 文件夹下的项目源程序文件夹,依次单击"New"→"Package"选项,新建文件夹 fragment,用于对源程序的分类管理。

(2)在 package Explore 视图中打开 src 文件夹,右击"fragment"文件夹,依次单击"New"→"Java Class"选项,在"Create New Class"对话框中(见图 5-13)的"Name"处填写"PersonalCenter",在"Superclass"中输入"android.app.Fragment",设置完成后单击"OK"按钮。

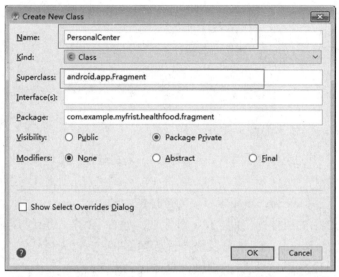

图 5-13 "Create New Class"对话框

（3）打开 PersonalCenter.java 文件，使用快捷键 Ctrl+O 打开"Select Methods to Override/Implement"对话框（见图 5-14），在该对话框中选中 onCreateView()方法，单击"OK"按钮，就可以在 PersonalCenter.java 文件中添加 onCreateView()方法。

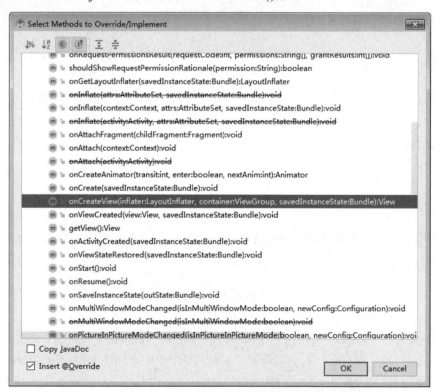

图 5-14 "Select Methods to Override/Implement"对话框

（4）重写 PersonalCenter.java 文件中的 onCreateView()方法，具体代码及其功能说明如下。

1.　public View onCreateView(LayoutInflater inflater, ViewGroup container, Bundle savedInstanceState) {

```
2.          //利用布局加载器加载"个人中心"布局，将其转换为 View
3.          View view = inflater.inflate(R.layout.frag_personalcenter, null);
4.          return view; //返回 view
5.      }
```

5.6 将"个人中心"碎片组装至 App 主框架

1. 知识点

Fragment 动态加载方法。

2. 工作任务

将创建完成的"个人中心"碎片组装至"良心食品"App 主框架中，如图 5-15 所示。组装完成后，在单击 App"底部导航"栏中的"个人中心"选项卡（图 5-15 中序号 1 处）时，能够将"个人中心"碎片在 App 内部进行显示（图 5-15 中序号 2 处）。

图 5-15 "个人中心"组装效果图

3. 操作流程

（1）在 package Explore 视图打开项目 src 文件夹中的 MainActivity.java 文件，修改该文件中的 onCreate()方法，如图 5-16 所示，添加标注处的代码。

```
public class MainActivity extends Activity {
    private RadioGroup myradio;
    private RadioButton rbutton1,rbutton2,rbutton3,rbutton4;
    private Resources res;
    private Drawable icon_home_true, icon_home_false,icon_me_true,icon_me_false,icon_order_true;
    private int fontColor_false,fontColor_true;
    private FragmentManager fgm;

    @Override
    protected void onCreate(Bundle savedInstanceState) {
        super.onCreate(savedInstanceState);
        this.requestWindowFeature(Window.FEATURE_NO_TITLE);
        setContentView(R.layout.activity_main);
        fgm = this.getFragmentManager();    ← 获取Fragment管理器
        intview();
        navigation();
    }
```

图 5-16 onCreate()方法代码

（2）修改 navigation()方法，如图 5-17 所示，分别添加标注的 3 行代码。这 3 行代码的作用如下。

代码行 1：FragmentTransaction transaction = fgm.beginTransaction()用于开启一个碎片管理事务。

代码行 2：transaction.replace(R.id.main_framelayout,new PersonalCenter())用于在布局中替换"个人中心"碎片。

代码行 3：transaction.commit()用于提交碎片管理事务。

```
private void navigation() {
    myradioGroup.setOnCheckedChangeListener((radioGroup, i) -> {
        FragmentTransaction transaction = fgm.beginTransaction();   → 1
        switch (i){
            case R.id.main_radio0://当"首页"选项被选中时，设置按钮选中状态时的文字及图片
                rbutton1.setTextColor(fontColor_true);
                rbutton1.setCompoundDrawablesWithIntrinsicBounds(left: null, icon_home_true, right: null, bottom: null);
                Toast.makeText( context: MainActivity.this, text: "首页", Toast.LENGTH_SHORT).show();
                break;
            case R.id.main_radio1://当"吃货驾到"选项被选中时，设置按钮选中状态时的文字及图片
                rbutton2.setTextColor(fontColor_true);
                rbutton2.setCompoundDrawablesWithIntrinsicBounds(left: null, icon_community_true, right: null, bottom: null);
                Toast.makeText( context: MainActivity.this, text: "吃货驾到", Toast.LENGTH_SHORT).show();
                break;
            case R.id.main_radio2://当"我的订单"选项被选中时，设置按钮选中状态时的文字及图片
                rbutton3.setTextColor(fontColor_true);
                rbutton3.setCompoundDrawablesWithIntrinsicBounds(left: null, icon_order_true, right: null, bottom: null);
                Toast.makeText( context: MainActivity.this, text: "我的订单", Toast.LENGTH_SHORT).show();
                break;
            case R.id.main_radio3://当"个人中心"选项被选中时，设置按钮选中状态时的文字及图片
                rbutton4.setTextColor(fontColor_true);
                rbutton4.setCompoundDrawablesWithIntrinsicBounds(left: null, icon_me_true, right: null, bottom: null);
                transaction.replace(R.id.main_framelayout, new PersonalCenter());   → 2
                Toast.makeText( context: MainActivity.this, text: "个人中心", Toast.LENGTH_SHORT).show();
                break;
        }
        transaction.commit();   → 3
```

图 5-17 navigation()方法代码

5.7 实现登录界面的调用

1. 知识点

在 Fragment 控件中添加监听方法。

2. 工作任务

本次任务主要实现首次运行 App 时，用户单击"个人中心"界面（见图 5-18）中的"待君登录"文本，跳转到"登录"界面（见图 5-19）。

图5-18　"个人中心"界面

图5-19　"登录"界面

3. 操作流程

（1）在 package Explore 视图中打开项目 src 文件夹中的 PersonalCenter.java 文件，并在该文件中创建 jumplogin()方法，具体代码如下。

```
1.   public void jumplogin(View v){
2.       TextView login=(TextView) v.findViewById(R.id.textView);
3.       login.setOnClickListener(new OnClickListener(){
4.           public void onClick(View v) {
5.               //创建显式 Intent，确认从当前 Activity 跳转到"登录"界面
6.               Intent it=new Intent(getActivity(),LoginActivity.class);
7.               startActivity(it);//启动 Activity
8.           }
9.       });
10.  }
```

（2）修改 onCreateView()方法，在该方法中调用 jumplogin()方法，具体代码如下。

```
1.   public View onCreateView(LayoutInflater inflater, ViewGroup container, Bundle savedInstanceState) {
2.       View view = inflater.inflate(R.layout.frag_cmember, null);
3.       jumplogin(view);//调用 jumplogin()方法，实现"登录"界面的跳转
4.       return view;
5.   }
```

第 6 章 "首页"模块的设计

教学目标

◆ 了解什么是适配器。
◆ 了解常用的适配器。
◆ 掌握 ArrayAdapter 的使用方法。
◆ 掌握 SimpleAdapter 的使用方法。
◆ 掌握 Spinner 组件的使用方法。
◆ 掌握 ListView 组件的使用方法。
◆ 掌握 GridView 组件的使用方法。
◆ 掌握 ViewPager 的使用方法。
◆ 掌握动态添加组件的方法。
◆ 掌握 LayoutParams 的使用方法。
◆ 掌握 UI 线程。
◆ 掌握 Handler 的使用方法。

 ## 6.1 工作任务概述

App 中不同的页面具有不同的作用。按照作用划分,可以将 App 页面大致分为 4 种类型:聚合类页面、列表类页面、内容类页面、功能类页面。聚合类页面多见于 App 的首页,用于功能入口的聚合展示。本章的主要工作任务是完成"良心食品"的"首页"页面制作,即完成如图 6-1 所示的效果,并实现以下功能。

(1)将"首页"页面加入 App 框架内,打开应用便显示该页面,单击"底部导航"的"首页"选项卡后加载并显示"首页"页面。

(2)实现"首页"顶部广告轮播效果(图 6-1 中标记 1 处)。

(3)实现"首页"中的六宫格 UI 效果(图 6-1 中标记 2 处),单击六宫格中的每个选项有相应的消息框显示功能(图 6-1 中标记 3 处)。

图 6-1 首页效果图

 6.2 预备知识

6.2.1 适配器

1. 适配器概述

适配器（Adapter）是数据和界面之间的桥梁。后台数据（如数组、链表、数据库、集合等）通过适配器变成手机页面能够正常显示的数据，可以理解为界面数据绑定，如果将数据、适配器和页面比作 MVC 模式，那么适配器在这中间充当 Controller 的角色。

一般来说，Spinner（下拉列表）、ListView（列表视图）、GridView（网格视图）、Gallery（画廊）、ViewPager 等组件都需要使用适配器来为其设置数据源。

android.widget.Adapter 类层次结构图如图 6-2 所示。

在图 6-2 中可以看到在 Android 中与适配器有关的所有接口、类的完整层级图，在使用过程中可以根据需求对接口或继承类进行相应扩展。比较常用的适配器有 BaseAdapter、ArrayAdapter、SimpleAdapter、SimpleCursorAdapter 等。

（1）BaseAdapter 是一个抽象类，继承它需要实现较多的方法，其具有较高的灵活性。

（2）ArrayAdapter 支持泛型操作，最为简单，只能展示一行文字。

（3）SimpleAdapter 的扩充性较好，可以通过自定义实现各种效果。

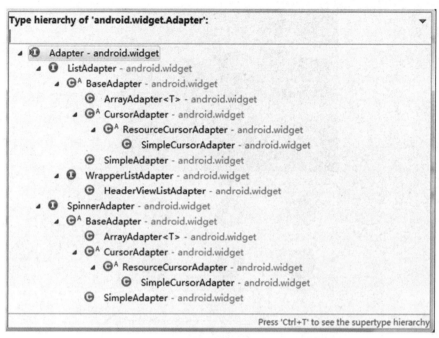

图 6-2　android.widget.Adapter 类层次结构图

（4）SimpleCursorAdapter 适用于简单的纯文字型 ListView，在使用时需要将 Cursor 的字段和 UI 的 ID 对应起来，若需要实现更复杂的 UI 也可以重写其他方法。SimpleCursorAdapter 可以理解为 SimpleAdapter 对数据库的简单整合，它可以方便地将数据库中的内容以列表的形式进行展示。

2. ArrayAdapter

ArrayAdapter 主要用于将简单的文本字符串在高级组件中进行显示。使用 ArrayAdapter 的步骤如下。

第一步：使用 new 运算符创建 ArrayAdapter 的对象。

例如：

```
ArrayAdapter arrayadapter = new ArrayAdapter(Context context,@layoutresource int resource, data);
```

第一个参数 Context context 是上下文，即当前视图所关联的且正在使用的适配器所处的上下文对象。第二个参数@layoutresource int resource 是 android sdk 中内置的一个布局，该布局中只有一个 TextView（该参数表明数组中每一条数据的布局是这个 View，即将每一条数据都显示在这个 View 上）。第三个参数 data 是要显示的数据。ArrayAdapter 既可以接收 List 作为数据源，也可以接收数组作为数据源。如果传入的是一个数组，那么 ArrayAdapter 会在构造函数中通过 Array.asList()方法将数组转换成 List。

第二步：组件调用 setAdapter()方法绑定适配器。

3. SimpleAdapter

SimpleAdapter 的扩展性较好，其可以定义各种各样的布局，可以适配 ImageView（图片）、Button（按钮）、CheckBox（复选框）等。

SimpleAdapter 的构造函数如下：

SimpleAdapter(Context context,List<?extends Map<String,?>>data,int resource,String[]from,int[]to)

（1）context：当前视图所关联的且正在使用的适配器所处的上下文对象。

（2）data：一个 Map 型列表，该列表中的每个条目对应列表中的一行。Map 中包含每一行的数据，并且包括所有的条目，data 可以理解为要装载的数据。

（3）resource：一个 View 布局的资源标记，其定义了布局中的列表项，布局文件中至少包含哪些需要展示的视图项、需要展示的布局样式。

（4）from：列名的列表，其在 Map 中对应每一项数据 item，是定义的 Map<String,Object>中的 String 数组。

（5）to：根据 from 参数可以得到的数值。to 对应的值是根据 from 参数的列表中的某个值得到的，是定义的 Map<String,Object>中的 Object 数组。

6.2.2 控件

1. Spinner

Spinner 提供了从一个数据集合中快速选择一项值的方法。默认情况下，Spinner 显示的是当前选择的值，单击 Spinner 会弹出一个包含所有可选值的 dropdown 菜单，可以在该菜单中为 Spinner 选择一个新值。如果开发者在使用 Spinner 时可以确定列表选择框里的列表项，则完全不需要编写代码，只需要先在 values/string.xml 文件中创建字符串数组，再将数组名指定给 entries 属性即可。如果程序需要在运行时动态确定 Spinner 的列表项，或者需要对 Spinner 的列表项进行定制，则可使用 Adapter 提供列表项。

（1）Spinner 的常用属性及其说明（见表 6-1）。

表 6-1 Spinner 的常用属性及其说明

属性	说明
android:entries	直接在XML布局文件中绑定数据源
android:spinnerMode	设置下拉框风格。android:spinnerMode="dropdown"为默认方式，是下拉列表风格；android:spinnerMode="dialog"为弹出对话框风格
android:popupBackground	设置下拉框背景色
android:dropDownHorizontalOffset	设置水平偏移量
android:dropDownVerticalOffset	设置垂直偏移量

（2）Spinner 的常用方法。

getSelectedItemPosition()：此方法用于获取用户在 Spinner 组件中选取的选项（该选项的索引编号从 0 开始）。

setOnItemSelectedListener()：此方法用于实现 OnItemSelectedListener 接口的监听对象。OnItemSelectedListener 接口有两个方法，具体代码如下。

```
1.    spinner.setOnItemSelectedListener(new AdapterView.OnItemSelectedListener(){
2.        public void onItemSelected(AdapterView<?> parent, View view, int position, long id) {
3.        }
```

```
4.      public void onNothingSelected(AdapterView<?> parent) {
5.          // TODO
6.      }
7.  });
8.
```

① onItemSelected()：当用户选择列表中的选项时会调用此方法。第三个参数 position 是常用的参数值，它保存了选中的 Spinner 中的列表项所在的位置值，一般自上而下编排，从 0 开始。

② onNothingSelected()：当用户拉下菜单但没有选取选项时会调用此方法。通常都不处理此方法，但因要实现接口中定义的所有方法，所以在定义监听器时仍要列出一个没有内容的 onNothingSelected()方法。

2. ListView

ListView（列表视图）是 Andorid 中常用的一种视图组件，它以垂直列表的形式列出需要显示的列表项。在 Android 中，有两种方法可以实现向屏幕中添加 ListView：一种是直接使用 ListView 组件；另一种是利用 Activity 继承 ListActivity。ListView 组件在使用时需要配合适配器将数据显示在视图上。

（1）ListView 的常用属性及其说明（见表 6-2）。

表 6-2 ListView 的常用属性及其说明

属　　性	说　　明
android:divider	分割线颜色。例如：android:divider="#f9b68b"
android:dividerHeight	分割线边距。例如：android:dividerHeight="1dp"
android:scrollbars	是否显示滚动条。例如：android:scrollbars="none"
android:listSelector	选中时的颜色，默认为橙黄底色。例如：android:listSelector="@color/pink"
android:transcriptMode	用ListView或其他显示大量Items的控件实时跟踪或查看信息，以实现最新的条目可以自动滚动到可视范围内。通过设置控件的transcriptMode属性可以使Android平台的控件（支持ScrollBar）自动滑动到底部。例如：android:transcriptMode="alwaysScroll"
android:footerDividersEnabled	当设置为false时，ListView将不会在各个footer之间绘制divider，默认为true。例如：android:footerDividersEnabled="false"
android:headerDividersEnabled	当设为false时，ListView将不会在各个header之间绘制divider，默认为true。例如：android:headerDividersEnabled ="false"

（2）ListView 的常用方法。

void addFooterView(View v)：增加一个固定在列表底部的 View，参数 v 为欲添加的视图。

void addFooterView(View v,Object data,boolean isSelectable)：增加一个固定在列表底部的 View，参数 v 为欲添加的视图，参数 data 为与 View 绑定的数据，参数 isSelectable 设置是否可选。

boolean removeFooterView(View v)：去除一个之前添加的 FooterView，参数 v 为欲删除的视图，若成功删除则返回 true。

void addHeaderView(View v)：增加一个固定在列表顶部的 View，参数 v 为欲添加的视图。

void addHeaderView(View v,Object data,boolean isSelectable)：增加一个固定在列表顶部的 View，参数 v 为欲添加的视图，参数 data 为与 View 绑定的数据，参数 isSelectable 设置是否可选。

boolean removeHeaderView(View v)：去除一个之前添加的 HeaderView，参数 v 为欲删除的视图，若成功删除则返回 true。

setOnItemClickListener()：添加选项单击监听事件。虽然 ListView 的用法和 Spinner 的用法非常相似，但 ListView 的默认行为没有选取事件。用户单击列表选项触发的是单击事件，而非选取事件，要监听此事件，必须使用 setOnItemClickListener()方法。

3. GridView

GridView（网格视图）可以将屏幕上的多个元素（文字、图片或其他组件）按网格的排列方式全部显示出来，在实现相册、图片浏览等应用时非常方便。在实现 GridView 时，需要用与 SimpleAdapter 类似的适配器来适配需要显示的元素（此时允许用户对其中的某一个元素进行操作），同时需要设置事件监听器 onItemClickListener 来捕捉和处理事件。

（1）GridView 的常用属性及其说明（见表 6-3）。

表 6-3 GridView 的常用属性及其说明

属性	说明
android:checkedButton	子单选按钮应该在默认情况下对其所在单选按钮组进行检查的ID
android:contentDescription	定义文本简要描述的视图内容
android:orientation	单选按钮排列的方式
android:numColumns	列数
android:columnWidth	每列的宽度
android:verticalSpacing	垂直间距，行间距
android:horizontalSpacing	水平间距，列间距
android:stretchMode	缩放模式
android:cacheColorHint="#00000000"	去除拖动时默认的黑色背景（常用属性）
android:listSelector="#00000000"	去除选中时的黄色底色（常用属性）
android:scrollbars="none"	隐藏GridView的滚动条
android:fadeScrollbars="true"	设置为true可以实现滚动条的自动隐藏和自动显示
android:fastScrollEnabled="true"	GridView出现快速滚动的按钮（至少滚动4页才会显示）
android:fadingEdge="none"	GridView衰落（褪去）时边缘颜色为空，默认值是vertical（可以理解为上下边缘的提示色）

（2）GridView 的常用方法。

```
1.    setOnItemClickListener(new OnItemClickListener(){
2.        public void onItemClick(AdapterView<?> parent, View view, int position, long id) {
3.    }
```

```
4.    });
```

onItemClick()方法中的第三个参数 int 类型的 position 返回的是 GridView 中被单击的网格的索引，索引从 0 开始。

4. ViewPager

ViewPager 可以使视图滑动，用于实现多页面切换的效果，它位于 Google 的兼容包 android-support-v4.jar 中。ViewPager 具有如下特点。

- 由于 ViewPager 类直接继承了 ViewGroup 类，因此它是一个容器类，可以在其中添加其他的 View 类。
- ViewPager 类需要 PagerAdapter 适配器类为它提供数据。
- ViewPager 通常与 Fragment 结合使用。Android 提供专门的 FragmentPagerAdapter 类和 FragmentStatePagerAdapter 类供 Fragment 中的 ViewPager 使用。

（1）ViewPager 的常用方法。

- setAdapter()：设置适配器。
- getCurrentItem()：设置当前选中页面。
- getCurrentItem()：获取当前选中页面角标。
- addOnPageChangeListener(mOnPageChangeListener)：为 ViewPager 添加页面改变的监听。
- removeOnPageChangeListener(mOnPageChangeListener)：移除 ViewPager 页面改变的监听。
- clearOnPageChangeListeners()方法：清除 ViewPager 所有的页面监听。

（2）ViewPager 的使用步骤。

第一步，在布局文件中添加一个 ViewPager 控件。

```
1.    <android.support.v4.view.ViewPager
2.        android:id="@+id/mViewPager"
3.        android:layout_width="wrap_content"
4.        android:layout_height="wrap_content">
5.    </android.support.v4.view.ViewPager>
```

第二步，在代码中找到该控件。

```
mViewPager =(ViewPager)findViewById(R.id.mViewPager);
```

第三步，新建一个类继承 PagerAdapter 类，并重写 PagerAdapter 类中的 getCount()方法、isViewFromObject()方法、instantiateItem()方法、destoryItem()方法。

```
1.    class MyAdapter extends PagerAdapter{
2.        //显示的个数
3.        @Override
4.        public int getCount () {
5.            return imageId.length;
6.        }
7.        //显示的 View 是否是当前的 View
8.        @Override
9.        public boolean isViewFromObject (View view, Object object) {
10.           return view == object;
11.       }
```

```
12.        //添加条目
13.        @Override
14.        public Object instantiateItem (ViewGroup container, int position) {
15.            Log.d (TAG, "instantiateItem: "+position);
16.            ImageView iv = new ImageView (getApplicationContext ());
17.            iv.setImageResource (imageId[position]);
18.            //向容器中添加一个 View
19.            container.addView (iv);
20.            return iv;
21.        }
22.        //销毁条目
23.        @Override
24.        public void destroyItem (ViewGroup container, int position, Object object) {
25.            Log.d (TAG, "destroyItem: "+position);
26.            //从容器中删除一个 View
27.            container.removeView ((View)object);
28.        }
29.    }
```

第四步，创建 MyAdapter 对象。

```
adapter = new MyAdapter();
```

第五步，通过 setAdapter()方法为 ViewPager 设置 MyAdapter 对象。

```
mViewPager.setAdapter(adapter);
```

5. LayoutParams

LayoutParams 继承自 Android.View.ViewGroup.LayoutParams，其相当于 Layout 的信息包，封装了 Layout 的位置、高、宽等信息。假设屏幕中的一块区域是由一个 Layout 占领的，如果将一个 View 添加到这个 Layout 中，最好给出 Layout 用户期望的布局方式，也就是将一个 Layout 用户认可的 LayoutParams 传递到这个 Layout 中。LayoutParams 类用于 child view（子视图）向其 parent view（父视图）传达自己的意愿（可以理解为孩子向其父亲说明自己想变成什么样）。

LayoutParams 有以下几个特点。

（1）LayoutParams 是一个 ViewGroup 的内部类，它属于基类，主要描述了宽与高。宽与高有三种指定方式。

① FILL_PARENT(renamed MATCH_PARENT in API Level 8 and higher)：填充父窗体。

② WRAP_CONTENT：包裹内容。

③ an exact number：精准描述。

（2）每一个继承自 ViewGroup 的容器都有其对应的 LayoutParams，并且这些 LayoutParams 又有各自独特的属性。

（3）子控件在获取 LayoutParams 时一定要和当前父控件的容器类型保持一致。若 TextView 是在 LinearLayout 下面的，那么 LayoutParams 必须是 LinearLayout.LayoutParams。

下面是一个 LayoutParams 的案例。

（1）UI 布局代码（Activity_main.XML）。

```xml
1.  <?xml version="1.0" encoding="utf-8"?>
2.  <LinearLayout xmlns:android="http://schemas.android.com/apk/res/android"
3.      xmlns:app="http://schemas.android.com/apk/res-auto"
4.      xmlns:tools="http://schemas.android.com/tools"
5.      android:layout_width="match_parent"
6.      android:layout_height="match_parent"
7.      android:orientation="vertical"
8.      android:id="@+id/Root"
9.      tools:context=".MainActivity">
10.     <TextView
11.         android:layout_width="100dp"
12.         android:layout_height="30dp"
13.         android:text="参照物"
14.         android:gravity="center"
15.         android:background="#6f00" />
16. </LinearLayout>
```

（2）动态添加组件代码（MainActivity.java）。

```java
1.  public class MainActivity extends Activity {
2.      private LinearLayout mRootView;
3.      private LinearLayout mLinearLayout;
4.      protected void onCreate(Bundle savedInstanceState) {
5.          super.onCreate(savedInstanceState);
6.          setContentView(R.layout.activity_main);
7.          //将 LinearLayout 添加到布局中
8.          //为 LinearLayout 新建一个 LinearLayout.LayoutParams，并且通过新建的 LinearLayout.LayoutParams（etLayoutParams）使 Layout 更新
9.          // LayoutParams 可以从父窗体获得，也可以自己创建，这里采用自己创建的方式
10.         mLinearLayout = new LinearLayout(this);//新建一个 LinearLayout（线性布局）
11.         mLinearLayout.setBackgroundColor(Color.parseColor("#0000ff"));
12.         mRootView=this.findViewById(R.id.Root);
13.         //新建一个 LayoutParams
14.         LinearLayout.LayoutParams layoutParams = new LinearLayout.LayoutParams(LinearLayout.LayoutParams.MATCH_PARENT,LinearLayout.LayoutParams.WRAP_CONTENT);
15.         //将新建的 LayoutParams 应用于新创建的线性布局
16.         mLinearLayout.setLayoutParams(layoutParams);
17.         mRootView.addView(mLinearLayout);
18.         mLinearLayout.setGravity(Gravity.CENTER);
19.         //将 TextView 添加到第 10 行代码创建的线性布局 mLinearLayout 中
20.         TextView textView = new TextView(this);
21.         textView.setText("新添加");
22.         textView.setBackgroundColor(Color.parseColor("#ff0000"));
23.         textView.setGravity(Gravity.CENTER);
24.         mLinearLayout.addView(textView);
25.         //为 TextView 获取对应父窗体类型的 LayoutParams 并设置参数更新 Layout
26.         LinearLayout.LayoutParams textParams = new LinearLayout.LayoutParams(textView.getLayoutParams());
27.         textParams.width = 200;
28.         textParams.height = 200;
29.         textView.setLayoutParams(textParams);
```

30.   }
31.   }

（3）本案例的 UI 布局效果如图 6-3（a）所示；利用动态添加组件的方法和 LayoutParams 设置 UI 布局的相关属性，效果如图 6-3（b）所示。

（a）

（b）

图 6-3　LayoutParams 案例效果图

6. 线程

（1）与 Android 相关的线程简介。

一个 Android 应用程序默认只有一个进程，但是一个进程可以有多个线程。其中的一个 UI 线程，即 UI Thread（主线程），在 Android 应用程序开始运行时就被创建，该线程主要负责控制 UI 界面的显示、更新和控件交互。所有的 Android 应用程序组件（包括 Activity、Service、Broadcast Receiver）都在应用程序的主线程中运行。任何组件中的费时操作都可能阻塞其他组件的运行，包括 Service 和可见的 Activity。主线程如果长时间无法响应，将出现 ANR（应用程序无响应）的错误，为了避免 ANR，耗时操作一般都开启子线程处理。

（2）线程的常用方法。

start():void：启动一个线程，系统会开启一个新线程来执行用户写的任务，并分配相应的资源。

run():void：当通过 start()方法启动了一个线程后，该线程获得 CPU 执行时间后则会自动进入 run()方法。因此要重写 thread()方法必须重写 run()方法，在 run()方法中定义要执行的任务。

thread.sleep(long millis):void：sleep()方法可以使线程睡眠，交出 CPU 以执行其他任务。但是 sleep()方法不会释放 monitor，如果当前线程持有某个对象，即使该线程进入睡眠状态，其他线程依旧无法访问该对象。因此使用 sleep()方法需要捕获异常。

interrupt():void：通过 interrupt()方法可以中断处于阻塞状态的线程。

isInterrupted( ):Boolean：判断一个线程是否被中断。interrupt()方法和 isInterrupted()方法结合使用可以中断处于非阻塞状态的线程。

（3）创建线程。

在 Java 中创建线程有两种方法：使用 Thread 类和使用 Runnable 接口。在使用 Runnable 接口时需要建立一个 Thread 实例。无论是使用 Thread 类还是使用 Runnable 接口建立线程，都必须建立 Thread 类或其子类的实例。

方法一：使用 Thread 类重写 run()方法。

```
1.   public class ThreadDemo1 {
2.       public static void main(String[] args){
3.           Demo d = new Demo();//创建线程实例
4.           d.start();//开启线程
5.       }
6.   }
7.   class Demo extends Thread{
8.       public void run(){
9.               System.out.println("我是 demo1 的线程");
10.          }
11.  }
```

方法二：使用 Runnable 接口重写 run()方法。

```
1.   public class ThreadDemo2 {
2.       public static void main(String[] args){
3.           Demo2 d =new Demo2();
4.           Thread t = new Thread(d);
5.           t.start();
6.           }
7.   }
8.   class Demo2 implements Runnable{
9.       public void run(){
10.          System.out.println("我是 demo2 的线程");
11.      }
12.  }
```

7. Handler

Android 应用程序在启动时，首先会开启一个主线程（UI 线程），负责管理界面中的 UI 控件。如果某线程操作 5s 还未完成，系统会发出错误提示。因此对于处理一些需要耗时的操作（如联网读取数据、读取本地较大的数据等操作），一般会将这些操作放在一个子线程中，但子线程不能修改 UI。Handler 则可以解决主线程与子线程通信的问题，Handler 运行在主线程（UI 线程）中，它与子线程可以通过 Message 对象来传递数据，从而达到主线程与子线程通信的目的。

（1）简单的 Handler 的使用步骤。

第一步：在程序中创建一个 Handler 对象，并且重写其 handleMessage()方法。

第二步：在创建的工作线程中创建一个 Message 对象，设置其响应的属性。

第三步：使用第一步创建的 Handler 对象将第二步创建的 Message 对象发送到主线程的消息队列中。

第四步：主线程中的 Looper 对象会不断地监听消息队列中的内容，当监听到消息队列中有新的内容时，就会从消息队列中获取此消息，并将该消息交给发送这个消息的 Handler 对象进行处理。

（2）handler 的常用方法。

handMessage()：用于处理消息。通过重写 handMessage()方法，处理从其他地方发送过来的消息。

sendMessage(message)：用于发送消息。

6.3 热身任务

6.3.1 "谁是你心中的英雄"

1. 任务说明

利用 Spinner+ArrayAdpater 完成如图 6-4（a）所示的 UI 效果；当单击下拉列表时产生如图 6-4（b）所示效果；当用户单击列表中的某个选项时出现消息框"你选择了：+选项"内容，如图 6-4（c）所示。

（a）

（b）

（c）

图 6-4 "谁是你心中的英雄"效果图

2. 操作步骤

(1) 创建一个 Android 项目。

(2) 在布局文件中添加相应的组件,完成如图 6-5 所示的布局效果。"谁是你心中的英雄"的 Component Tree 如图 6-6 所示。

图 6-5　初始布局　　　　　图 6-6　"谁是你心中的英雄"的 Component Tree

(3) 打开 MainActivity.java 源程序,在 onCreate()方法中输入以下内容。

```
1.   public class MainActivity extends Activity {
2.       private String[] hero_name={"毛泽东","马云","曹操","自己"};//在数组中定义下拉列表中显示的数据
3.       private Spinner myspinner;
4.       protected void onCreate(Bundle savedInstanceState) {
5.           super.onCreate(savedInstanceState);
6.           setContentView(R.layout.activity_main);
7.           myspinner=this.findViewById(R.id.spinner);
8.   //创建数组适配器,第一个参数是上下文,第二个参数用来规范下拉列表中每个选项的布局,第三个参数用于设置在下拉列表中显示的数据
9.           ArrayAdapter adapter=new ArrayAdapter(this,android.R.layout.simple_list_item_1,hero_name);
10.          myspinner.setAdapter(adapter);//将创建的数组适配器与下拉列表组件进行绑定
11.          myspinner.setOnItemSelectedListener(new AdapterView.OnItemSelectedListener() {
12.              public void onItemSelected(AdapterView<?> adapterView,View view,int i,long l) {
13.   //将用户选择项消息框中进行显示,其中 i 值保存了用户选择项的序号,序号从 0 开始编排
14.                  Toast.makeText(MainActivity.this,"你选择了:"+hero_name[i],Toast.LENGTH_LONG).show();
                 }
15.              public void onNothingSelected(AdapterView<?> adapterView) {
16.              }
17.          });
18.      }
19.  }
```

6.3.2 "永不消失的经典"

1. 任务说明

利用 GridView+SimpleAdpater 完成如图 6-7（a）所示的九宫格效果；当单击任一选项时则弹出消息框，消息框的内容为"你选择了:+标题"，如图 6-7（b）所示。

（a）

（b）

图 6-7 "永不消失的经典"效果图

2. 操作步骤

（1）创建一个 Android 项目。

（2）在布局文件中添加 GridView 组件，初始布局效果如图 6-8 所示。布局文件代码如下。

```
1.  <?xml version="1.0" encoding="utf-8"?>
2.  <android.support.constraint.ConstraintLayout
        xmlns:android="http://schemas.android.com/apk/res/android"
3.      xmlns:app="http://schemas.android.com/apk/res-auto"
4.      xmlns:tools="http://schemas.android.com/tools"
5.      android:layout_width="match_parent"
6.      android:layout_height="match_parent"
7.      tools:context=".MainActivity">
8.      <GridView
9.          android:id="@+id/gridView1"
10.         android:layout_width="match_parent"
11.         android:layout_height="match_parent"
12.         android:numColumns="3"
13.         app:layout_constraintStart_toStartOf="parent"
14.         app:layout_constraintTop_toTopOf="parent" />
15. </android.support.constraint.ConstraintLayout>
```

图 6-8 初始布局效果

（3）在项目中的\res\layout 文件夹中添加 itemview.xml 文件，该文件用于规范每个选项元素的界面布局。itemview.xml 文件代码如下。

```xml
1. <?xml version="1.0" encoding="utf-8"?>
2. <LinearLayout xmlns:android="http://schemas.android.com/apk/res/android"
3.     android:layout_width="match_parent"
4.     android:layout_height="match_parent"
5.     android:orientation="vertical"
6.     android:gravity="center">
7.     <ImageView
8.         android:id="@+id/imageView1"
9.         android:layout_width="100dp"
10.        android:layout_height="120dp"/>
11.    <TextView
12.        android:id="@+id/textView1"
13.        android:layout_width="wrap_content"
14.        android:layout_height="wrap_content"
15.        android:textSize="18sp"
16.        android:layout_marginTop="10dp"
17.        android:text="TextView" />
18. </LinearLayout>
```

（4）打开 MainActivity.java 源程序，重写 onCreate( )方法，实现 GridView 组件的数据适配及完成相应单击，具体代码及相关功能说明如下。

```java
1. public class MainActivity extends Activity {
2.     private String[] name={"地道战","董存瑞","51 号兵站","红色娘子军","建军大业","狼牙山五壮士",
           "建党伟业","青春之歌","中华女儿"};//定义数组，用于存储电影标题
3.     private int[] image={ R.drawable.p1_di, R.drawable.p2_dong, R.drawable.p3_five,
           R.drawable.p4_hong, R.drawable.p5_jianjun, R.drawable.p6_lang, R.drawable.p7_jiandang,
           R.drawable.p8_qing, R.drawable.p9_zhong};//定义数组，用于存储电影海报图片
4.     List imagelist;
5.     private GridView myGridView;
```

```
6.    protected void onCreate(Bundle savedInstanceState) {
7.        super.onCreate(savedInstanceState);
8.        setContentView(R.layout.activity_main);
9.        myGridView=(GridView) this.findViewById(R.id.gridView1);
10.       imagelist=new ArrayList();
11.       for(int i=0;i<9;i++){
12.           HashMap hm=new HashMap();
13.           hm.put("name", name[i]);
14.           hm.put("image", image[i]);
15.           imagelist.add(hm);
16.       }
17.       //创建简单适配器,第一个参数是上下文,第二个参数是用于显示图片和标题的数据集,第三个参数是用于规范每个选项显示样式的自定义布局,第四个参数是指定 HashMap 中哪些数据显示于 UI,第五个参数是指定将数据显示于自定义布局的哪个组件上。此处将 image、name 这两个数据分别显示于 UI 中的 imageView1、textView1 这两个组件上
18.       SimpleAdapter myAdapter=new SimpleAdapter(this, imagelist, R.layout.itemview, new String[]{"image","name"}, new int[]{R.id.imageView1,R.id.textView1});
19.       myGridView.setAdapter(myAdapter);//将适配器应用于 GridView 组件
20.       myGridView.setOnItemClickListener(new AdapterView.OnItemClickListener() {
21.           @Override
22.           public void onItemClick(AdapterView<?> adapterView,View view,int i,long l) {
23.               //当用户单击选项时用消息框显示信息,其中 i 值保存了被选中选项的序号,序号从 0 开始编排
24.               Toast.makeText(MainActivity.this,"你选择了:"+name[i],Toast.LENGTH_LONG).show();
25.           }
26.       });
27.   }
28. }
```

6.3.3 "我激动,我数数"

1. 任务说明

本任务在完成如图 6-9 所示的效果的同时完成以下两项功能。

图 6-9 "我激动,我数数"效果图

功能一：单击"开始数数"按钮，则计数文本标签从 0 开始，以每秒累加 1 的方式显示计数效果。

功能二：单击"停止数数"按钮，则停止计数功能。

2. 操作步骤

（1）创建一个 Android 项目。

（2）在布局文件中添加相应组件，制作如图 6-9 所示的布局效果。布局文件代码如下。

```
1.  <?xml version="1.0" encoding="utf-8"?>
2.  <LinearLayout xmlns:android="http://schemas.android.com/apk/res/android"
3.      xmlns:app="http://schemas.android.com/apk/res-auto"
4.      xmlns:tools="http://schemas.android.com/tools"
5.      android:layout_width="match_parent"
6.      android:layout_height="match_parent"
7.      android:orientation="vertical"
8.      tools:context=".MainActivity">
9.      <TextView
10.         android:id="@+id/count"
11.         android:layout_width="match_parent"
12.         android:layout_height="wrap_content"
13.         android:textColor="#ff0000"
14.         android:gravity="center"
15.         android:textSize="40sp"
16.         android:text="0" />
17.     <ImageView
18.         android:id="@+id/imageView"
19.         android:layout_width="match_parent"
20.         android:layout_height="wrap_content"
21.         android:src="@drawable/counter" />
22.     <Button
23.         android:id="@+id/start"
24.         android:layout_width="match_parent"
25.         android:layout_height="wrap_content"
26.         android:text="开始数数" />
27.     <Button
28.         android:id="@+id/end"
29.         android:layout_width="match_parent"
30.         android:layout_height="wrap_content"
31.         android:text="停止数数" />
32. </LinearLayout>
```

（3）打开 MainActivity.java 源程序，添加代码，实现相应功能，具体代码如下。

```
1.  public class MainActivity extends Activity implements View.OnClickListener {
2.      private Button bt_start, bt_end;
3.      private TextView tv_count;
4.      MyThread myThread;
5.      private int count = 0;
```

```java
6.      Handler handler;
7.      @SuppressLint("HandlerLeak")
8.      protected void onCreate(Bundle savedInstanceState) {
9.          super.onCreate(savedInstanceState);
10.         setContentView(R.layout.activity_main);
11.         //子线程每发送过来一个信息,Handler 就将 TextView 的文本数字累加一次并更新
12.         handler = new Handler() {
13.             public void handleMessage(Message msg) {
14.                 switch (msg.what) {
15.                     case 1:
16.                         count++;//用于显示的数字加 1
17.                         tv_count.setText(count + "");//修改 UI 组件属性
18.                         break;
19.                 }
20.             }
21.         };
22.         init();
23.     }
24.     public void init() {
25.         bt_start = (Button) findViewById(R.id.start);
26.         bt_end = (Button) findViewById(R.id.end);
27.         tv_count = (TextView) findViewById(R.id.count);
28.         bt_start.setOnClickListener(this);
29.         bt_end.setOnClickListener(this);
30.     }
31.     public void onClick(View view) {
32.         switch (view.getId()) {
33.             //若单击的是"开始数数"按钮,则开启子线程
34.             case R.id.start:
35.                 myThread = new MyThread();
36.                 myThread.start();//开启线程
37.                 Toast.makeText(this,"start",Toast.LENGTH_SHORT).show();
38.                 break;
39.             //若单击的是"停止数数"按钮,则中断子线程
40.             case R.id.end:
41.                 myThread.interrupt();//中断线程
42.                 Toast.makeText(this,"end",Toast.LENGTH_SHORT).show();
43.                 break;
44.         }
45.     }
46.     public class MyThread extends Thread {
47.         public void run() {
48.             //只要线程不中断则子线程每隔 1s 发送一个信息 1 给主线程的 Handler 进行 UI 修改操作
49.             while (!Thread.interrupted()) {
50.                 try {
51.                     Thread.sleep(1000);
52.                     handler.sendEmptyMessage(1);
53.                 } catch (InterruptedException e) {
54.                     e.printStackTrace();
55.                     break;
56.                 }
```

```
57.            }
58.          }
59.        }
60.  }
```

第 12～21 行代码创建了一个 handler，用于实现主线程与子线程的通信。

第 24～30 行代码创建了一个 init()方法，用于初始化工作。需要注意的是第 28 行及第 29 行代码是另一种添加监听的方法，这种方法多应用于对多个组件添加同一个监听事件，其使用要求是在类定义时通过 implements 实现监听接口（如第 1 行代码所示），同时要实现监听接口所要求的方法（如第 31～45 行代码所示）。

第 46～57 行代码定义一个内部类 MyThread 继承 Thread 类，并重写 run()方法实现每隔 1s 向 Handler 发送一次信息。

 小贴士

当使用了 sleep()方法、同步锁的 wait()方法、socket 的 receiver()方法、accept()方法等方法时，会使线程处于阻塞状态。当调用线程的 interrupt()方法时，系统会抛出一个 InterruptedException 异常，代码通过捕获异常，然后通过 break 跳出循环状态，使线程正常结束。若要正常结束 run()方法，一定要先捕获 InterruptedException 异常之后通过 break 来跳出循环。

6.3.4 "找不同"

1. 任务说明

本任务主要利用 ViewPager 控件实现两张图的左右滑动切换。"找不同"效果图如图 6-10 所示。

（a）　　　　　　　　　　　　　　（b）

图 6-10 "找不同"效果图

2. 操作步骤

（1）创建一个 Android 项目。

（2）将 diff1.png 及 diff2.png 两张图片复制到项目中的/app/res/drawable 文件夹中。

（3）在布局文件中添加 ViewPager。布局文件代码如下。

```
1.  ?xml version="1.0" encoding="utf-8"?>
2.  <android.support.constraint.ConstraintLayout
        xmlns:android="http://schemas.android.com/apk/res/android"
3.      xmlns:app="http://schemas.android.com/apk/res-auto"
4.      xmlns:tools="http://schemas.android.com/tools"
5.      android:layout_width="match_parent"
6.      android:layout_height="match_parent"
7.      tools:context=".MainActivity">
8.  <android.support.v4.view.ViewPager
9.      android:id="@+id/ViewPager1"
10.     android:layout_width="match_parent"
11.     android:layout_height="match_parent"
12.     app:layout_constraintTop_toTopOf="parent"
13.     app:layout_constraintStart_toStartOf="parent">
14. </android.support.v4.view.ViewPager>
15. </android.support.constraint.ConstraintLayout>
```

第 8~14 行代码实现了插入一个 ViewPager 组件的功能，由于该组件是 V4 兼容包中的组件，因此不能直接从组件面板拖曳至布局视图，只能通过手动输入代码的方式添加。

（4）打开 MainActivity.java 源程序并对其进行修改，具体代码如下。

```
1.  public class MainActivity extends Activity {
2.      private ViewPager viewPager;
3.      private ImageView imageView;
4.      private List<View> views;//定义变量 views，用于存放显示至 ViewPager 的 ImageView 组件
5.      private int imagelist[] = { R.drawable.diff1, R.drawable.diff2 };//定义 int 数组，用于存放图片
6.      protected void onCreate(Bundle savedInstanceState) {
7.          super.onCreate(savedInstanceState);
8.          setContentView(R.layout.activity_main);
9.          InitImageView();
10.         InitViewPager();
11.     }
12.     private void InitImageView() {
13.         views = new ArrayList(); //新建一个 ArrayList 对象，用于存放要显示的 ImageView
14.         for (int i = 0; i <= 1; i++) {
15.             ImageView Image = new ImageView(this);//新建一个 ImageView 对象
16.             image.setBackgroundResource(imagelist[i]);//设置图片框的背景图片
17.             views.add(image);//将创建好的 ImageView 添加到 List 中
18.         }
19.     }
20.     private void InitViewPager() {
21.         viewPager = (ViewPager) this.findViewById(R.id.ViewPager1);//查找 ViewPager 组件
22.         viewPager.setAdapter(new myAdapter());//设置 ViewPager 适配器为自定义的 PagerAdapter
23.     }
```

```
24.     private class myAdapter extends PagerAdapter {
25.         public int getCount() {
26.             return views.size();//ViewPager 要显示的视图个数
27.         }
28.         public boolean isViewFromObject(View arg0, Object arg1) {
29.             // TODO Auto-generated method stub
30.             return arg0 == arg1;//判断显示的 View 是否为当前的 View
31.         }
32.         //销毁条目
33.         public void destroyItem(ViewGroup container,int position,Object object) {
34.             // TODO Auto-generated method stub
35.             container.removeView(views.get(position));
36.         }
37.         public int getItemPosition(Object object) {
38.             // TODO Auto-generated method stub
39.             return super.getItemPosition(object);
40.         }
41.         //添加条目
42.         public Object instantiateItem(ViewGroup container, int position) {
43.             container.addView(views.get(position));
44.             return views.get(position);
45.         }
46.     }
47. }
```

第 12～19 行的 InitImageView()方法用于将要显示的两张图片放于 ImageView 中后添加至 List 中。

第 20～23 行的 InitViewPager()方法用于将要显示的两张图片适配到 ViewPager 组件，其适配器使用的是第 24～46 行代码创建的内部类。

第 24～46 行代码是一个继承 PagerAdapter 的内部类，用于创建一个 ViewPager 的适配器，实现将图片在 ViewPager 内进行显示。

6.4 创建"首页"Fragment

1. 知识点

➢ ViewPager 组件的添加方法。
➢ GridView 的常用属性。

2. 工作任务

制作"首页"模块的 UI 布局并创建相应的 Fragment。

3. 操作流程

（1）依次单击"File"→"New"→"XML"→"Layout XML File"选项，创建 frag_home.xml 文件。

（2）在 frag_home.xml 文件中添加组件。"首页"的 Component Tree 如图 6-11 所示。

图 6-11　"首页"的 Component Tree

（3）打开 frag_home.xml 文件，修改各组件的相关属性，具体代码如下。

```
1.    ?xml version="1.0" encoding="utf-8"?>
2.    <LinearLayout xmlns:android="http://schemas.android.com/apk/res/android"
3.        android:layout_width="match_parent"
4.        android:layout_height="match_parent"
5.        android:background=""
6.        android:orientation="vertical">
7.        <android.support.v4.view.ViewPager
8.            android:id="@+id/home_viewpager"
9.            android:layout_width="match_parent"
10.           android:layout_marginTop="5dp"
11.           android:layout_height="200dp">
12.       </android.support.v4.view.ViewPager>
13.       <LinearLayout
14.           android:id="@+id/home_viewpager_dot"
15.           android:layout_width="match_parent"
16.           android:layout_height="wrap_content"
17.           android:layout_alignParentBottom="true"
18.           android:background="#fff"
19.           android:gravity="center"
20.           android:orientation="horizontal"
21.           android:paddingBottom="10dp"
22.           android:paddingTop="10dp">
23.       </LinearLayout>
24.       <GridView
25.           android:id="@+id/home_gridView1"
26.           android:layout_width="match_parent"
27.           android:layout_height="wrap_content"
28.           android:layout_marginTop="8dp"
29.           android:background="#fff"
30.           android:cacheColorHint="#00000000"
31.           android:horizontalSpacing="1dp"
32.           android:listSelector="#fff"
33.           android:numColumns="3"
34.           android:paddingBottom="5dp"
35.           android:paddingTop="5dp"
36.           android:verticalSpacing="10dp" />
37.   </LinearLayout>
```

（4）在项目的 src\fragment 文件夹中新建 Home_fragment.java 继承 Fragment，同时添加 onCreateView()方法。

（5）重写 Home_fragment.java 中的 onCreateView()方法，具体代码及其功能说明如下。

1. public View onCreateView(LayoutInflater inflater, ViewGroup container,Bundle savedInstanceState) {
2. 　　　// TODO Auto-generated method stub
3. 　　　View view = inflater.inflate(R.layout.frag_home, null); //利用布局加载器加载"个人中心"布局，将其转换为 View
4. 　　　return view; //返回 View
5. }

6.5 将"首页"碎片组装至 App 主框架

1. 知识点

Fragment 动态加载方法。

2. 工作任务

将创建好的"首页"碎片（此时碎片页面内容为空白）组装至"良心食品"App 主框架中，如图 6-12 所示。组装完成后，单击 App"底部导航"栏中的"首页"选项卡（图 6-12 中标记 2 处）时，能够将"首页"碎片在 App 内进行显示（图 6-12 中标记 1 处）。

图 6-12 "首页"碎片组装效果图

3. 操作流程

（1）在 package Explore 视图中打开项目 src 文件夹中的 MainActivity.java 源程序，修改 initview()方法，用于打开 App 时的初始状态显示"首页"，如图 6-13 所示，在原有程序代码的基础上添加以下代码。

```
FragmentTransaction     transaction = fgm.beginTransaction();//开启 Fragment 事务
transaction.replace(R.id.main_framelayout, new Home_fragment());//替换碎片内容
transaction.commit();//提交事务
```

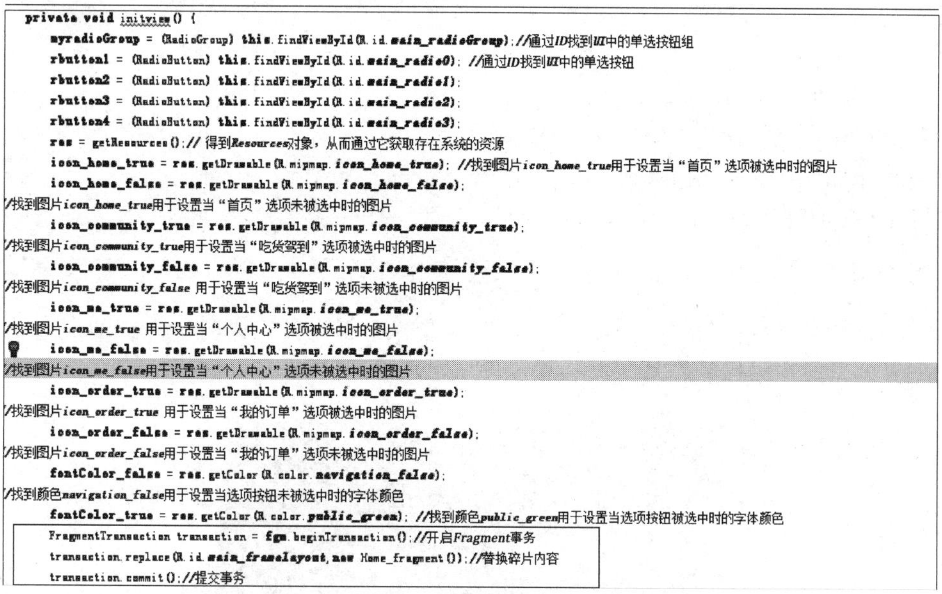

```
private void initview() {
    myradioGroup = (RadioGroup) this.findViewById(R.id.main_radioGroup);//通过ID找到UI中的单选按钮组
    rbutton1 = (RadioButton) this.findViewById(R.id.main_radio0); //通过ID找到UI中的单选按钮
    rbutton2 = (RadioButton) this.findViewById(R.id.main_radio1);
    rbutton3 = (RadioButton) this.findViewById(R.id.main_radio2);
    rbutton4 = (RadioButton) this.findViewById(R.id.main_radio3);
    res = getResources();// 得到Resources对象，从而通过它获取存在系统的资源
    icon_home_true = res.getDrawable(R.mipmap.icon_home_true); //找到图片icon_home_true用于设置当“首页”选项被选中时的图片
    icon_home_false = res.getDrawable(R.mipmap.icon_home_false);
//找到图片icon_home_true用于设置当“首页”选项未被选中时的图片
    icon_community_true = res.getDrawable(R.mipmap.icon_community_true);
//找到图片icon_community_true用于设置当“吃货驾到”选项被选中时的图片
    icon_community_false = res.getDrawable(R.mipmap.icon_community_false);
//找到图片icon_community_false 用于设置当“吃货驾到”选项未被选中时的图片
    icon_me_true = res.getDrawable(R.mipmap.icon_me_true);
//找到图片icon_me_true 用于设置当“个人中心”选项被选中时的图片
    icon_me_false = res.getDrawable(R.mipmap.icon_me_false);
//找到图片icon_me_false用于设置当“个人中心”选项未被选中时的图片
    icon_order_true = res.getDrawable(R.mipmap.icon_order_true);
//找到图片icon_order_true 用于设置当“我的订单”选项被选中时的图片
    icon_order_false = res.getDrawable(R.mipmap.icon_order_false);
//找到图片icon_order_false用于设置当“我的订单”选项未被选中时的图片
    fontColor_false = res.getColor(R.color.navigation_false);
//找到颜色navigation_false用于设置当选项按钮未被选中时的字体颜色
    fontColor_true = res.getColor(R.color.public_green); //找到颜色public_green用于设置当选项按钮被选中时的字体颜色
    FragmentTransaction transaction = fgm.beginTransaction();//开启Fragment事务
    transaction.replace(R.id.main_framelayout,new Home_fragment());//替换碎片内容
    transaction.commit();//提交事务
```

图 6-13 initview 方法代码

（2）修改 navigation()方法，用于当切换至“导航”中的“首页”选项卡时将“首页”模块加载至程序主框架中，如图 6-14 所示，在原有程序代码的基础上添加以下代码。


```
transaction.replace(R.id.main_framelayout, new Home_fragment());
```

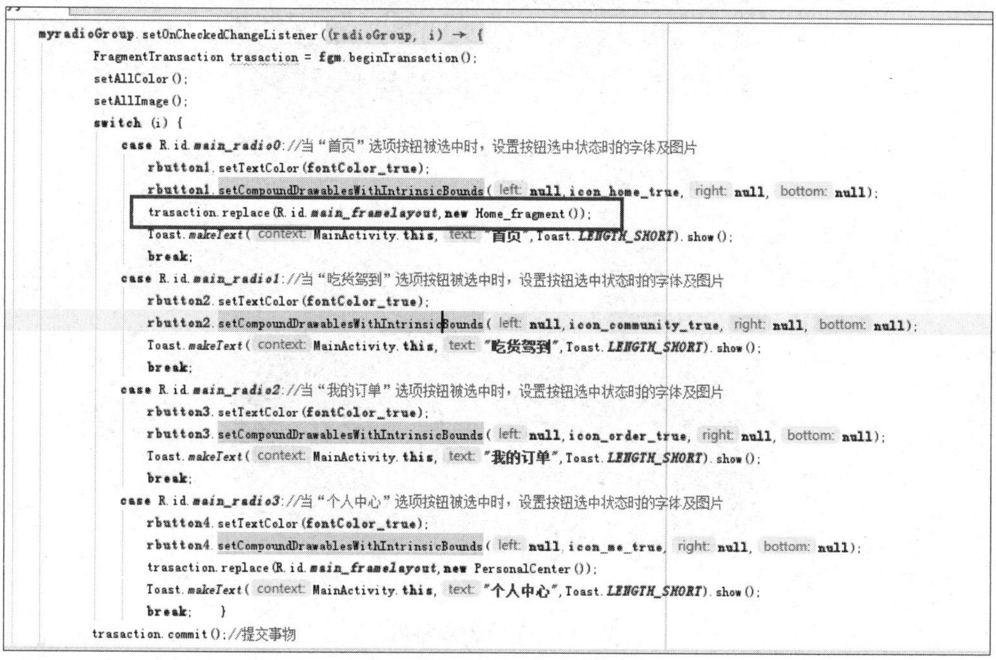

```
myradioGroup.setOnCheckedChangeListener((radioGroup, i) -> {
        FragmentTransaction trasaction = fgm.beginTransaction();
        setAllColor();
        setAllImage();
        switch (i) {
            case R.id.main_radio0://当“首页”选项按钮被选中时，设置按钮选中状态时的字体及图片
                rbutton1.setTextColor(fontColor_true);
                rbutton1.setCompoundDrawablesWithIntrinsicBounds( left: null,icon_home_true, right: null, bottom: null);
                trasaction.replace(R.id.main_framelayout,new Home_fragment());
                Toast.makeText( context: MainActivity.this, text: "首页",Toast.LENGTH_SHORT).show();
                break;
            case R.id.main_radio1://当“吃货驾到”选项按钮被选中时，设置按钮选中状态时的字体及图片
                rbutton2.setTextColor(fontColor_true);
                rbutton2.setCompoundDrawablesWithIntrinsicBounds( left: null,icon_community_true, right: null, bottom: null);
                Toast.makeText( context: MainActivity.this, text: "吃货驾到",Toast.LENGTH_SHORT).show();
                break;
            case R.id.main_radio2://当“我的订单”选项按钮被选中时，设置按钮选中状态时的字体及图片
                rbutton3.setTextColor(fontColor_true);
                rbutton3.setCompoundDrawablesWithIntrinsicBounds( left: null,icon_order_true, right: null, bottom: null);
                Toast.makeText( context: MainActivity.this, text: "我的订单",Toast.LENGTH_SHORT).show();
                break;
            case R.id.main_radio3://当“个人中心”选项按钮被选中时，设置按钮选中状态时的字体及图片
                rbutton4.setTextColor(fontColor_true);
                rbutton4.setCompoundDrawablesWithIntrinsicBounds( left: null,icon_me_true, right: null, bottom: null);
                trasaction.replace(R.id.main_framelayout,new PersonalCenter());
                Toast.makeText( context: MainActivity.this, text: "个人中心",Toast.LENGTH_SHORT).show();
                break;   }
        trasaction.commit();//提交事物
```

图 6-14 navigation 方法代码

6.6 实现"首页"图片轮播效果

1. 知识点

   - ViewPager 的使用方法。
   - PagerAdapter 的使用方法。
   - LayoutParams 的使用方法。
   - Handler 的使用方法。

2. 工作任务

在"首页"顶部实现图片轮播效果，每隔 5s 更换一次图片，同时下面的圆点指示器在相应位置改变。任务完成后的图片轮播效果如图 6-15～图 6-17 所示。

图 6-15 图片轮播效果一

图 6-16 图片轮播效果二

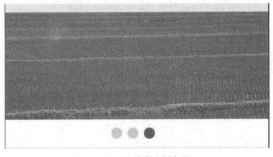

图 6-17 图片轮播效果三

3. 操作流程

（1）在 java 文件夹中新建一个 Adapter 文件夹，用于存放自定义适配器。

（2）在 java/Adapter 文件夹中新建一个用于适配 ViewPager 图片的 HomeVpAdapter.java 类，该类用于继承 PagerAdapter，并重写 getCount()等方法，具体代码及相关功能说明如下。

```java
1.  public class HomeVpAdapter extends PagerAdapter{
2.      List<ImageView> mImages;//此变量用于存放轮播图片
3.      //此方法用于在实例化时将图片以参数的形式传入适配器
4.      public HomeVpAdapter(List<ImageView> images){
5.          this.mImages=images;
6.      }
7.      //ViewPager 切换的页面数
8.      public int getCount() {
9.          // TODO Auto-generated method stub
10.         return mImages.size();
11.     }
12.     //判断显示的 View 是否为当前的 View
13.     public boolean isViewFromObject(View view, Object object) {
14.         // TODO Auto-generated method stub
15.         return view == object;
16.     }
17.     //销毁视图
18.     public void destroyItem(ViewGroup container, int position, Object object) {
19.         // TODO Auto-generated method stub
20.         container.removeView((View) object);
21.     }
22.     //组装显示视图
23.     public Object instantiateItem(ViewGroup container, int position) {
24.         // TODO Auto-generated method stub
25.         container.addView(mImages.get(position));
26.         return mImages.get(position);
27.     }
28. }
```

（3）在项目的 drawable 文件夹中依次单击"New"→"Android Resource File"选项，新建一个名为 don_false_shape.xml 的 shape 文件（该文件用于设置轮播图中指示器未被选中的小圆点），并在该文件中添加相应属性，具体代码如下。

```xml
1.  <?xml version="1.0" encoding="utf-8"?>
2.  <shape xmlns:android="http://schemas.android.com/apk/res/android"
3.      android:shape="oval"
4.      android:useLevel="false">
5.      //此处引用了第 4 章在 color.xml 文件中定义的颜色
6.      <solid android:color="@color/navigation_false"/>
7.      <size android:width="10dp"
8.          android:height="10dp"/>
9.  </shape>
```

（4）在项目的 drawable 文件夹中依次单击"New"→"Android Resource File"选项，新建一

个名为 don_true_shape.xml 的 shape 文件（该文件用于设置轮播图中指示器被选中的小圆点），并在该文件中添加相应属性，具体代码如下。

```xml
1.  <?xml version="1.0" encoding="utf-8"?>
3.  <shape xmlns:android="http://schemas.android.com/apk/res/android"
3.      android:shape="oval"
4.      android:useLevel="false">
5.      //此处引用了第 4 章在 color.xml 文件中定义的颜色
6.      <solid android:color="@color/public_green"/>
7.      <size android:width="10dp"
8.          android:height="10dp"/>
9.  </shape>
```

（5）在项目的 drawable 文件夹中依次单击"New"→"Android Resource File"选项，新建一个名为 tv_dot_selector.xml 的 selector 文件（该文件用于实现切换指示器小圆点的选中与未选中指示状态），并在该文件中添加相应属性，具体代码如下。

```xml
1.  <?xml version="1.0" encoding="utf-8"?>
2.  <selector xmlns:android="http://schemas.android.com/apk/res/android" >
3.      <item android:state_selected="true"android:drawable="@drawable/don_true_shape"></item>
4.      <item android:state_selected="false" android:drawable="@drawable/don_false_shape"></item>
5.  </selector>
```

（6）打开 Home_fragment.java 程序，在该程序中添加代码以实现图片动态轮播及圆点指示器指示功能，具体代码及相关功能说明如下。

```java
1.  public class Home_fragment extends Fragment {
2.      private ViewPager viewpager;//轮播图组件
3.      private List<ImageView> list_VP = new ArrayList<ImageView>();//定义 List，用于存放 ImageView
4.      private int[] Image_VP = {R.drawable.guanggao1,R.drawable.guanggao2,R.drawable.guanggao3};//此数组存放轮播图片
5.      TextView tv_radio;//用于显示指示器小圆点
6.      LinearLayout ll_viewpager_don;//此线型布局用于动态加载指示器小圆点
7.      //定义此 List，用于存放 3 个指示器小圆点
8.      private List<TextView> tv_list = new ArrayList<TextView>();
9.      //用于设定动态加载的小圆点指示器在布局中的位置
10.     private LinearLayout.LayoutParams layoutParams;
11.     Thread mytr;//定义一个子线程，用于耗时处理
12.     private Handler handler = new Handler() {
13.         public void handleMessage(Message msg) {
14.             //在当前页面位置加 1
15.             switch(msg.what){
16.                 case 1:
17.                     //利用取模运算方法（模为 3）实现动态切换 ViewPager 图片的功能
18.                     viewpager.setCurrentItem((viewpager.getCurrentItem() + 1)%3);
19.                 }
20.             }
21.     };
22.     public View onCreateView(LayoutInflater inflater,@Nullable ViewGroup container,Bundle savedInstanceState) {
23.         //利用布局加载器加载"个人中心"布局，将其转换为 View
```

```java
24.        View view = inflater.inflate(R.layout.frag_home,null);
25.        initViewPager(view);
26.        return view; //返回 View
27.    }
28.    private void initViewPager(View view) {
29.        viewpager = (ViewPager) view.findViewById(R.id.home_viewpager);
30.        ll_viewpager_don = (LinearLayout) view
31.                .findViewById(R.id.home_viewpager_dot);//此线型布局用于动态加载指示器小圆点
32.        //定义一个 LayoutParams 属性,用于设置子控件相对于父控件的大小为宽 20、高 20
33.        layoutParams = new LinearLayout.LayoutParams(20,20);
34.        layoutParams.setMargins(0,0,10,0);//设置 LayoutParams 的 Margins 值
35.        for (int i = 0; i < Image_VP.length; i++) {
36.            //加入轮播图片
37.            ImageView imageV = new ImageView(getActivity());
38.            imageV.setImageResource(Image_VP[i]);
39.            imageV.setAdjustViewBounds(true);//设置能够调整图片的边框
40.            //不按比例缩放图片,图片塞满整个 ImageView
41.            imageV.setScaleType(ImageView.ScaleType.FIT_XY);
42.            list_VP.add(imageV);
43.            //新建一个 TextView,注意这里的上下文是通过 getActivity()方法获取的,此方法是在一个 Fragment 中获取上下文的常用方法
44.            tv_radio = new TextView(getActivity());
45.            tv_radio.setBackgroundResource(R.drawable.tv_dot_selector);//设置 TextView 背景
46.            ll_viewpager_don.addView(tv_radio,layoutParams);//将 TextView 添加至界面指示器
47.            tv_list.add(tv_radio);//将 TextView 添加至 List,用于更改其显示状态
48.        }
49.        viewpager.setAdapter(new HomeVpAdapter(list_VP));
50.        viewpager.setCurrentItem(0); //程序开始,将 ViewPager 的显示索引值设置为 0
51.        tv_list.get(0).setSelected(true); //程序开始,将第 1 个小圆点设置为选中状态
52.        //监听 ViewPager 滚动,主要用于设置圆点指示器的位置及显示状态
53.        viewpager.setOnPageChangeListener(new ViewPager.OnPageChangeListener() {
54.            @Override
55.            public void onPageScrolled(int position,float positionOffset,int positionOffsetPixels) {
56.            }
57.            public void onPageSelected(int position) {
58.                for (int i = 0; i < tv_list.size(); i++) {
59.                    if (i == position % list_VP.size()) {
60.                        tv_list.get(i).setSelected(true);
61.                    } else {
62.                        tv_list.get(i).setSelected(false);
63.                    }
64.                }
65.            }
66.            public void onPageScrollStateChanged(int state) {
67.            }
68.        });
69.    }
70.    private void myThread(){
71.        mytr=new Thread(new Runnable(){
```

```
72.        public void run() {
73.            // TODO Auto-generated method stub
74.            while(!Thread.interrupted()){
75.                try {
76.                    Thread.sleep(5000);
77.                    handler.sendEmptyMessage(1);//每隔 5s 向 Handler 发送一次消息
78.                } catch (InterruptedException e) {
79.                    // TODO Auto-generated catch block
80.                    e.printStackTrace();
81.                }
82.            }
83.        }
84.    });
85.    mytr.start();
86. }
87. public void onResume() {
88.     myThread();
89.     super.onResume();
90. }
91. @Override
92. public void onStop() {
93.     mytr.interrupt();//结束当前 Fragment 时中断耗时的子线程
94.     super.onStop();
95. }
96. }
```

上述代码包括 1 个 Handler 匿名内部类和 5 个方法，即 onCreateView()方法、initViewPager()方法、myThread()方法、onResume()方法、onStop()方法。其中，onCreateView()方法、onResume()方法、onStop()方法是 Fragment 生命周期中的 3 个方法，initViewPager()方法、myThread()方法是自定义方法。

第 24～62 行代码中的 initViewPager()方法的功能是完成 ViewPager 数据的适配和指示器的动态添加。需要注意的是，在第 46～62 行代码中，监听 ViewPager 的 OnPageChangeListener 事件的目的是，每当 ViewPager 页面改变就让指示器中的 3 个小圆点更新一次选中状态，从而实现指示器功能。

第 80～83 行代码重写 onResume()方法是为了开启计时子线程。

第 85～88 行代码重写 onStop()方法是为了结束 Fragment 时结束计时子线程。

第 62～79 行代码组成的 myThread()方法的功能是实现每隔 5s 向主线程的 Handler 队列发送一次信息，而主线程的 Handler 类（第 10～18 行代码）接收到信息后切换 ViewPager 的显示页。

 思考

结合 Fragment 的生命周期思考为什么要在 onResume( )方法中开启子线程而不是在 onCreateView( )方法中开启子线程。

6.7 实现"首页"的数据适配功能

1. 知识点

➢ GridView 数据适配方法。
➢ GridView 的 OnItemClickListener 监听方法。
➢ SimpleAdapter 的使用方法。

2. 工作任务

这里的工作任务是实现"首页"数据适配的功能。当单击六宫格中的每个选项时显示相应的标题信息,如单击"新品驾到",则在消息框中显示"新品驾到",效果如图 6-18 所示。

图 6-18 单击"新品驾到"效果图

3. 操作流程

(1) 在项目的\res\layout 文件夹中添加 buju_home_gridview.xml 文件,该文件用于规范 GridView 每个选项的布局样式。buju_home_gridview.xml 文件代码如下。

```
1.   <?xml version="1.0" encoding="utf-8"?>
2.   <LinearLayout xmlns:android="http://schemas.android.com/apk/res/android"
3.       android:id="@+id/one_buju_relativelayout"
4.       android:layout_width="match_parent"
5.       android:layout_height="match_parent"
6.       android:gravity="center"
7.       android:orientation="vertical" >
8.       <ImageView
9.           android:id="@+id/buju_home_gridview_icon"
10.          android:layout_width="50dp"
11.          android:layout_height="50dp"/>
12.      <TextView
13.          android:id="@+id/buju_home_gridview_name"
14.          android:layout_width="wrap_content"
15.          android:layout_height="wrap_content"
16.          android:layout_marginTop="5dp"
```

```
17.            android:textColor="#999999"
18.            android:textSize="13sp" />
19.    </LinearLayout>
```

（2）打开 Home_fragment.java 程序，在该程序中新建 initGridView()方法用于实现 GridView 的数据适配及 OnItemClickListener 监听的添加，具体代码及相关功能说明如下。

```
1.    private void initGridView(View view) {
2.        gridView = (GridView) view.findViewById(R.id.home_gridView1);
3.        //创建动态数组，用于存放 HashMap 数据
4.        list = new ArrayList<HashMap<String, Object>>();
5.        for (int i = 0; i < 6; i++) {
6.            HashMap<String, Object> hmap = new HashMap<String, Object>();
7.            hmap.put("name",name_listview[i]);
8.            hmap.put("icon",image_listview[i]);
9.            list.add(hmap);
10.       }
11.       //创建简单适配器，用于 GridView 数据适配
12.       SimpleAdapter adapter = new SimpleAdapter(getActivity(),list, R.layout.buju_home_gridview, new String[]{"name","icon"},new int[]{R.id.buju_home_gridview_name,R.id.buju_home_gridview_icon});
13.       //创建简单适配器，为数据适配到 GridView 组件做准备
14.       gridView.setAdapter(adapter);
15.       //为 GridView 添加 OnItemClickListener 监听，实现用户单击"GridView"选项时弹出消息框
16.       gridView.setOnItemClickListener(new AdapterView.OnItemClickListener() {
17.           @SuppressLint("WrongConstant")
18.           public void onItemClick(AdapterView<?> arg0,View v,int i,long l) {
19.               switch (i) {
20.                   case 0:
21.                       Toast.makeText(getActivity(),"新品驾到",5000).show();
22.                       break;
23.                   case 1:
24.                       Toast.makeText(getActivity(),"食趣",5000).show();
25.                       break;
26.                   case 2:
27.                       Toast.makeText(getActivity(),"食品安全",5000).show();
28.                       break;
29.                   case 3:
30.                       Toast.makeText(getActivity(),"产品溯源",5000).show();
31.                       break;
32.                   case 4:
33.                       Toast.makeText(getActivity(),"健康养身",5000).show();
34.                       break;
35.                   case 5:
36.                       Toast.makeText(getActivity(),"产品展示",5000).show();
37.                       break;
38.               }
39.           }
40.       });
41.   }
```

第 2 行代码中的 gridView 在 Home_fragment.java 程序中以成员变量形式声明为 private

GridView gridView。

第 4 行代码中的 list 在 Home_fragment.java 程序中以成员变量形式声明为 private List list。

第 7 行代码中的 name_listview 在 Home_fragment.java 程序中以成员变量形式声明为 private String name_listview[] = {"新品驾到","食趣","食品安全","产品溯源","健康养生","产品展示"}。

第 8 行代码中的 image_listview[]在 Home_fragment.java 程序中以成员变量形式声明为 private int image_listview[]={R.mipmap.no1,R.mipmap.no2,R.mipmap.no3,R.mipmap.no4,R.mipmap.no5,R.mipmap.no6}。

（3）在 onCreateView()方法中调用 initGridView()方法，具体代码如下。

```
1.    public View onCreateView(LayoutInflater inflater, ViewGroup container, Bundle savedInstanceState) {
2.        View view = inflater.inflate(R.layout.frag_home, null);
3.        initViewPager(view);
4.        initGridView(view);
5.        return view;
6.    }
```

第 7 章 "吃货驾到"模块的设计

教学目标

◇ 掌握 BaseAdapter 的使用方法。
◇ 掌握 ContextMenu 的使用方法。
◇ 掌握 AlertDialog 的使用方法。

7.1 工作任务概述

本章的主要工作任务是完成"吃货驾到"页面的制作,需要完成以下工作子任务。"吃货驾到"效果图如图 7-1 所示。

(1)完成"吃货驾到"页面的 UI 布局,效果如图 7-1(a)所示。

(a)　　　　　　　　(b)　　　　　　　　(c)

图 7-1　"吃货驾到"效果图

（2）制作"吃货驾到"页面的 Fragment。

（3）将"吃货驾到"页面添加到有"底部导航"的主 Activity 框架内（第 4 章创建的 MainActivity）。

（4）在"吃货驾到"页面中，单击图中标记 1 所示的心形图标（点赞），能够将点赞数显示于图中标记 2 处，如图 7-1（b）所示。

（5）为"吃货驾到"页面的每个分享内容创建上下文菜单，效果如图 7-1（c）所示。

 ## 7.2 预备知识

7.2.1 BaseAdapter

1. BaseAdapter 概述

BaseAdapter 是 Android 应用程序中十分常用的基础数据适配器，是非常实用的一个类。相比第 6 章介绍的 ArrayAdapter、SimpleAdapter 两类适配器，BaseAdapter 较难理解，但由于其具有全能性，因此 Android 应用程序开发者必须掌握 BaseAdapter 的使用方法。

2. BaseAdapter 的使用方法

使用 BaseAdapter 的目的主要是通过继承此类来实现 BaseAdapter 的如下 4 个方法。

（1）public int getCount()：获取适配器中数据集的数据个数。

（2）public Object getItem(int position)：获取数据集中与索引对应的数据项。

（3）public long getItemId(int position)：获取指定行对应的 ID。

（4）public View getView(int position, View convertView, ViewGroup parent)：获取每一行 Item 显示的内容。

BaseAdapter 先通过 getCount()方法确定数量，然后循环执行 getView()方法将条目逐一绘制出来，必须重写的是 getCount()方法和 getView()方法。而 getItem()方法和 getItemId()方法是调用某些函数时才会触发的方法，如果不使用，可以暂时不修改。以下代码为这 4 种方法的应用示例。

```
1.    public class myadapter extends BaseAdapter{
2.        //getCount()方法是程序在加载到 UI 上时就要先读取的,这里获得的值决定了 UI 组件显示的数据项数
3.        public int getCount() {
4.            // TODO Auto-generated method stub
5.            return 0;
6.        }
7.        //根据 UI 组件所在位置返回 View
8.        public Object getItem(int position) {
9.            return null;
10.       }
11.       //根据 UI 组件所在位置得到数据源集合中的 ID
12.       public long getItemId(int position) {
```

```
13.              return 0;
14.        }
15.        //获取每一行 Item 的显示内容
16.        public View getView(int position, View convertView, ViewGroup parent) {
17.              return null;
18.        }
19. }
```

BaseAdapter 常用的两个方法如下。

（1）notifyDataSetChanged()：提醒依附在监视器底层的数据已发生改变，每一个 Item 视图都应该刷新本身。

（2）notifyDataSetInvalidated()：提醒依附在监视器底层的数据不再是有效的或可获得的。

7.2.2 菜单

菜单是 Android 应用程序中非常重要且常见的组成部分，其主要可以分为 3 类：选项菜单、上下文菜单及弹出菜单，它们的主要区别如下。

（1）选项菜单（OptionsMenu）是一个应用程序的主菜单项，用于放置对应用程序产生全局影响的操作，如搜索、设置。

（2）上下文菜单（ContextMenu）是用户长按某个元素时出现的浮动菜单，它提供的操作将影响所选内容，主要应用于列表中的每项元素（如长按列表项弹出"删除"对话框）。

（3）弹出菜单（PopupMenu）以垂直列表形式显示一系列操作选项，一般由某个控件触发，显示在对应控件的上方或下方。弹出菜单用于提供与特定内容相关的大量操作。

7.2.3 ContextMenu

1. ContextMenu 概述

当用户对某个 View 长时间按住不放时，弹出的菜单称为 ContextMenu，这类菜单只能显示标题，不能显示图标。

2. ContextMenu 的使用方法

一般通过以下方法来使用 ContextMenu。

（1）创建上下文菜单。

```
onCreateContextMenu(ContextMenu menu,View v,ContextMenuInfo menuInfo);
```

该方法是一个回调函数，用于创建 ContextMenu，ContextMenu 每次显示时都会调用这个函数。其中，参数 v 指定上下文菜单绑定的 View，而 ContextMenuInfo 则是该上下文菜单的一些额外信息，额外信息包含以下字段。

① public long id：用于显示上下文菜单的子视图的行 ID。

② public int position：用于显示上下文菜单的子视图在适配器中的位置。

③ public View targetView：用于显示上下文菜单的子视图，是 AdapterView 的子视图之一。

④ public AdapterView.AdapterContextMenuInfo(View targetView,int position,long id)：构造

函数。

（2）响应菜单单击事件。

```
onContextItemSelected(MenuItem item);
```

一般会利用 item.getItemId()方法来获取被选中菜单选项的 ID。

（3）为 View 注册上下文菜单。

```
registerForContextMenu(View view);
```

该方法用于为某个 View 注册上下文菜单。

（4）为菜单添加菜单选项。

```
menu.add(int groupId,int itemId,int order,int titleRes);
```

该方法用于为上下文菜单添加菜单选项。其中，各参数说明如下。

① groupId：int 类型的 group ID 参数，代表的是组概念。该参数可以将几个菜单项归为一组，以便以组的方式管理菜单按钮。

② itemId：int 类型的 item ID 参数，代表的是项目编号。这个参数非常重要，一个 item ID 对应 Menu 中的一个选项。在后面使用菜单的时候，是靠这个 item ID 来判断使用的是哪个选项的。

③ order：int 类型的 order 参数，代表的是菜单项的显示顺序。默认是 0，表示菜单的显示顺序是按照 order 的显示顺序来显示的。

④ titleRes：String 类型的 title 参数，表示选项中显示的文字。

3. ContextMenu 的使用步骤

（1）在 Activity 或 Fragment 中调用 registerForContextMenu(View v)方法，注册需要和上下文菜单关联的 View。如果将 ListView 或 GridView 作为参数传入 ContextMenu，那么每个列表项将会有相同的浮动上下文菜单。

（2）在 Activity 或 Fragment 中重写 onCreateContextMenu()方法，加载 Menu 资源。

（3）在 Activity 或 Fragment 中重写 onContextItemSelected()方法，实现菜单项的单击逻辑。

7.2.4 对话框

1. 对话框概述

在 Android 的开发中，在界面弹出对话框（Dialog）是与用户进行交互的常用手段。对话框并不会占满整个屏幕，其通常用于某项事件当中，用户做出选择后才会继续执行。Android 中提供了多种类型的对话框以满足开发的需要：普通对话框、列表对话框、单选对话框、多选对话框、等待对话框、进度条对话框、自定义对话框，如图 7-2 所示。

2. 普通对话框

普通对话框是一种较常用的对话框，一般可以显示 1 个标题和最多 3 个按钮。在使用普通对话框时会用到几种常用的方法，具体如表 7-1 所示。

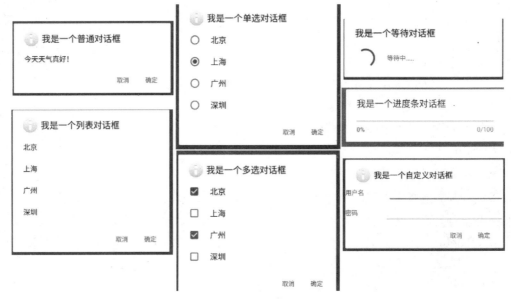

图 7-2　对话框效果图

表 7-1　普通对话框的常用方法及其功能

方　　法	功　　能
setTitle()	设置对话框标题
setIcon()	设置对话框图标
setPositiveButton()	设置对话框的"确定"按钮
setNegativeButton()	设置对话框的"取消"按钮
setMessage()	设置对话框提示信息

下面的代码展示了利用上述方法实现的一个简单的普通对话框。

```
1.   public class MainActivity extends Activity {
2.      protected void onCreate(Bundle savedInstanceState) {
3.          super.onCreate(savedInstanceState);
4.          setContentView(R.layout.activity_main);
5.          AlertDialog.Builder dialog=new AlertDialog.Builder(this);//创建普通对话框
6.          dialog.setTitle("你好");//设置对话框标题
7.          dialog.setIcon(android.R.drawable.ic_dialog_alert);//设置对话框图标
8.          dialog.setMessage("你好吗？");//设置对话框提示信息
9.          dialog.setPositiveButton("确定", null);//添加"确定"按钮
10.         dialog.setNegativeButton("取消", null) ;//添加"取消"按钮
11.         dialog.show();//显示对话框
12.     }
13. }
```

第 5 行代码中的 new AlertDialog.Builder() 方法用的参数是 this，表示的是上下文。

小贴士

（1）普通对话框通常用 show()方法显示对话框，而用 dismiss()方法关闭对话框。

（2）上述代码中仅创建了"取消"按钮及"确定"按钮，而未实现其单击功能，若要实现单击功能还需要在它们的第二个参数中添加单击监听事件，具体代码如下。

```
1.   dialog.setNegativeButton ("取消",new DialogInterface.OnClickListener() {
2.          public void onClick(DialogInterface dialogInterface,int i) {
3.          }
4.   });
```

（3）因为普通对话框的构造方法全部是 Protected 的，所以不能直接通过 new 创建普通对话框。若要创建普通对话框，需要使用 AlertDialog.Builder。

 7.3 热 身 任 务

本节热身任务为"通讯录"。

1. 任务说明

通过 ListView+BaseAdapter 实现如图 7-3 所示的 UI 效果。

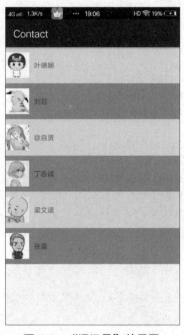

图 7-3 "通讯录"效果图

2. 操作步骤

（1）创建一个 Android 项目。

（2）打开 activity_main.xml 文件并完成布局效果，具体代码如下。

```
1.  <?xml version="1.0" encoding="utf-8"?>
2.  <LinearLayout xmlns:android="http://schemas.android.com/apk/res/android"
3.      xmlns:app="http://schemas.android.com/apk/res-auto"
4.      xmlns:tools="http://schemas.android.com/tools"
5.      android:layout_width="match_parent"
6.      android:layout_height="match_parent"
7.      android:orientation="vertical"
8.      tools:context=".MainActivity">
9.      <ListView
10.         android:id="@+id/contact"
11.         android:layout_width="match_parent"
12.         android:layout_height="match_parent" />
13. </LinearLayout>
```

(3) 将图片复制到项目的 drawable 文件夹中。

(4) 在项目的\res\layout 文件夹中添加 contect_item.xml 文件，该文件用于规范每个选项元素的界面布局。contect_item.xml 文件代码如下。

```
1.  <?xml version="1.0" encoding="utf-8"?>
2.  <LinearLayout xmlns:android="http://schemas.android.com/apk/res/android"
3.      xmlns:app="http://schemas.android.com/apk/res-auto"
4.      xmlns:tools="http://schemas.android.com/tools"
5.      android:layout_width="match_parent"
6.      android:gravity="center_vertical"
7.      android:layout_height="match_parent"
8.      android:paddingTop="10dp"
9.      android:paddingBottom="10dp">
10.     <ImageView
11.         android:id="@+id/contact_photo"
12.         android:layout_width="49dp"
13.         android:layout_height="52dp"   />
13.     <TextView
15.         android:id="@+id/contact_name"
16.         android:layout_width="wrap_content"
17.         android:layout_height="wrap_content"
18.         android:layout_marginLeft="10dp"
19.         android:text="TextView" />
20. </LinearLayout>
```

(5) 打开 java 文件夹下的 MainActivity.java 源程序，修改相关代码，实现相关功能，具体代码如下。

```
1.  public class MainActivity extends Activity {
2.      private ListView mylistview;
3.      private int images[] = {R.drawable.p1,R.drawable.p2,R.drawable.p3,R.drawable.p4,R.drawable.p5,R.drawable.p6};//联系人图像
4.      private String[] name = {"叶德娴","刘芸","徐自贤 ","丁志诚 ","梁文道 ","张笛 "};//联系人姓名
5.      private List<HashMap> dataList;//List 数据集用于存放联系人数据集
6.      protected void onCreate(Bundle savedInstanceState) {
7.          super.onCreate(savedInstanceState);
8.          setContentView(R.layout.activity_main);
```

```
9.          mylistview = (ListView) this.findViewById(R.id.contact);
10.         initdata();
11.         MyBaseAdapter adapter = new MyBaseAdapter();
12.         mylistview.setAdapter(adapter);
13.     }
14.     //初始化数据,将每个人的图像及姓名通过 HashMap 捆绑在一起,放在 List 中
15.     private void initdata() {
16.         dataList = new ArrayList();
17.         for (int i = 0; i < images.length; i++) {
18.             HashMap hm = new HashMap();
19.             hm.put("image",images[i]);
20.             hm.put("name",name[i]);
21.             dataList.add(hm);
22.         }
23.     }
24.     //定义内部类以继承 BaseAdapter,用于数据适配
25.     public class MyBaseAdapter extends BaseAdapter {
26.         //在组件中显示的数据个数
27.         public int getCount() {
28.             return dataList.size();
29.         }
30.         public Object getItem(int position) {
31.             return null;
32.         }
33.         public long getItemId(int position) {
34.             return 0;
35.         }
36.         //个性化地生成每一行 Item 的显示内容
37.         public View getView(int position,View convertView,ViewGroup parent) {
38.             LayoutInflater layInflater = LayoutInflater.from(MainActivity.this);
39.             View view = layInflater.inflate(R.layout.contect_item,null);
40.             ImageView image = (ImageView) view.findViewById(R.id.contact_photo);
41.             TextView name = (TextView) view.findViewById(R.id.contact_name);
42.             image.setBackgroundResource(Integer.parseInt(dataList.get(position).get("image").toString()));
43.             name.setText(dataList.get(position).get("name").toString());
44.             //设置偶数行背景为黄色,设置奇数行背景为绿色
45.             if (position % 2 == 0) {
46.                 view.setBackgroundColor(Color.parseColor("#FFFF00"));
47.             } else {
48.                 view.setBackgroundColor(Color.parseColor("#66CD00"));
49.             }
50.             return view;
51.         }
52.     }
53. }
```

第 15~23 行代码中的 initdata()方法的主要功能是将碎片数据(图像、名字)捆绑成一个对象便于后面引用。

第 25~51 行代码组成的 MyBaseAdapter 类是本程序的核心部分，继承自 BaseAdapter。与第 6 章介绍的 ArrayAdapter 及 SimpleAdapter 两种适配器相比，MyBaseAdapter 的使用更为复杂，需要重写其方法来规范数据的适配方法，但也因为如此，它更灵活，功能更强大，如本例的奇偶行背景色的设置就体现了其灵活与功能的强大。

7.4 创建"吃货驾到" Fragment

1. 知识点

➢ ListView 组件的添加方法。
➢ ListView 的常用属性。
➢ Fragment 的创建。

2. 工作任务

制作"吃货驾到"页面的 UI 布局并创建相应的 Fragment。

3. 操作流程

（1）依次单击"File"→"New"→"XML"→"Layout XML File"选项，创建 frag_gourmet.xml 文件。

（2）在 frag_gourmet.xml 文件中添加组件。"吃货驾到"的 Component Tree 如图 7-4 所示。

图 7-4 "吃货驾到"的 Component Tree

（3）打开 frag_gourmet.xml 文件，修改各组件的相关属性，具体代码如下。

```
1.   <?xml version="1.0" encoding="utf-8"?>
2.   <LinearLayout xmlns:android="http://schemas.android.com/apk/res/android"
3.       android:layout_width="match_parent"
4.       android:layout_height="match_parent"
5.       android:orientation="vertical">
6.       <ListView
7.           android:id="@+id/gourment_frag_listView1"
8.           android:layout_width="match_parent"
9.           android:layout_height="match_parent"
10.          android:divider="#EAEAEA"
11.          android:dividerHeight="2dp"
12.          android:listSelector="#EAEAEA"/>
13.  </LinearLayout>
```

第 10 行代码用于设置分割线颜色。
第 11 行代码用于设置分割线边距。

第 12 行代码用于设置 listView 的 item 选中时的颜色。

（4）在项目的 java\fragment 文件夹中新建 Gourmet_fragment.java 程序，以继承 Fragment，同时在该程序中添加 onCreateView()方法。

（5）重写 Gourmet_fragment.java 程序中的 onCreateView( )方法，具体代码及相关功能说明如下。

```
1.   public View onCreateView(LayoutInflater inflater, ViewGroup container,Bundle savedInstanceState) {
2.       // TODO Auto-generated method stub
3.       //利用布局加载器加载"吃货驾到"布局，将其转换为 View
4.       View view = inflater.inflate(R.layout.frag_gourmet, null);
5.       return view; //返回 View
6.   }
```

第 4 行代码利用布局加载器加载"吃货驾到"布局，并将其转换为 View。

第 5 行代码返回 View。

 ## 7.5　将"吃货驾到"碎片组装至 App 主框架

1. 知识点

Fragment 动态加载方法。

2. 工作任务

在 7.4 节工作任务的基础上将创建完成的"吃货驾到"碎片（此时碎片内容是空白的）组装至"良心食品"App 主框架中，如图 7-5 所示。组装完成后，在单击 App"底部导航"栏中的"吃货驾到"选项卡（标记 2 处）时，能够将"吃货驾到"碎片在 App 内（标记处）进行显示。

图 7-5　"吃货驾到"碎片组装效果图

3. 操作流程

（1）在 package Explore 视图下打开项目 java 文件夹中的 MainActivity.java 程序，修改 navigation()方法，用于当切换至"导航"栏中的"吃货驾到"选项卡时将"吃货驾到"页面加载至程序框架中，如图 7-6 所示，在原有程序代码的基础上添加以下代码。

```
trasaction.replace(R.id.main_framelayout,new Gourmet_fragment());
```

图 7-6 navigation()方法代码

7.6 实现"吃货驾到"的数据适配功能

1. 知识点

用 BaseAdapter 适配数据的方法。

2. 工作任务

将"吃货驾到"的数据适配显示到 listView 组件上，完成后的效果图如图 7-7 所示。

3. 操作流程

（1）在项目的 drawable 文件夹中依次单击"New"→"Android Resource File"选项，新建一个名为 thumb_up.xml 的 selector 文件，并在该文件中添加相应属性、创建点赞选择器，具体代码如下。

```
1.  <?xml version="1.0" encoding="utf-8"?>
2.  <selector xmlns:android="http://schemas.android.com/apk/res/android" >
3.      <item android:state_pressed="true" android:drawable="@drawable/heart2"/>
4.      <item android:state_pressed="false" android:drawable="@drawable/heart1"/>
5.  </selector>
```

图 7-7 "吃货驾到"数据适配效果图

第 3～4 行代码实现两种状态的切换。

(2) 在项目的\res\layout 文件夹中添加 buju_gourmet_listview.xml 文件,该文件用于规范每个选项元素 item 的界面布局。"吃货驾到"的 Component Tree 如图 7-8 所示。

图 7-8 "吃货驾到"的 Component Tree

buju_gourmet_listview.xml 文件代码如下。

```
1.  <?xml version="1.0" encoding="utf-8"?>
2.  <LinearLayout xmlns:android="http://schemas.android.com/apk/res/android"
3.      xmlns:app="http://schemas.android.com/apk/res-auto"
4.      xmlns:tools="http://schemas.android.com/tools"
5.      android:layout_width="match_parent"
6.      android:layout_height="match_parent"
7.      android:minHeight="100dp"
8.      android:orientation="horizontal">
```

```
9.      <ImageView
10.         android:id="@+id/buju_gourmet_pic"
11.         android:layout_width="100dp"
12.         android:layout_height="80dp"
13.         android:layout_marginTop="10dp"
14.         android:paddingLeft="30dp"
15.         android:scaleType="fitStart" />
16.     <LinearLayout
17.         android:layout_width="match_parent"
18.         android:layout_height="wrap_content"
19.         android:layout_marginLeft="10dp"
20.         android:layout_marginTop="15dp"
21.         android:orientation="vertical">
22.         <TextView
23.             android:id="@+id/buju_gourmet_text"
24.             android:layout_width="wrap_content"
25.             android:layout_height="wrap_content"
26.             android:layout_gravity="center_vertical"
27.             android:maxLines="5"
28.             android:text="内容"
29.             android:textColor="#000" />
30.         <android.support.constraint.ConstraintLayout
31.             android:layout_width="match_parent"
32.             android:layout_height="wrap_content"
33.             android:layout_marginTop="10dp">
35.             <TextView
36.                 android:id="@+id/buju_gourmet_username"
37.                 android:layout_width="wrap_content"
38.                 android:layout_height="wrap_content"
39.                 android:text="用户名"
40.                 android:textColor="#60000000"
41.                 android:textSize="12sp"
42.                 app:layout_constraintBottom_toBottomOf="parent"
43.                 app:layout_constraintStart_toStartOf="parent"
44.                 app:layout_constraintTop_toTopOf="parent" />
45.             <TextView
46.                 android:id="@+id/buju_gourmet_time"
47.                 android:layout_width="wrap_content"
48.                 android:layout_height="wrap_content"
49.                 android:layout_marginLeft="40dp"
50.                 android:text="时间"
51.                 android:textColor="#60000000"
52.                 android:textSize="12sp"
53.                 app:layout_constraintBottom_toBottomOf="parent"
54.                 app:layout_constraintLeft_toRightOf="@+id/buju_gourmet_username"
55.                 app:layout_constraintTop_toTopOf="parent" />
56.         </android.support.constraint.ConstraintLayout>
57.         <android.support.constraint.ConstraintLayout
58.             android:layout_width="match_parent"
59.             android:layout_height="wrap_content"
```

```
60.                android:layout_marginTop="10dp">
61.            <ImageView
62.                android:id="@+id/buju_gourment_praise"
63.                android:layout_width="15dp"
64.                android:layout_height="15dp"
65.                android:src="@drawable/thumb_up"
66.                app:layout_constraintBottom_toBottomOf="parent"
66.                app:layout_constraintStart_toStartOf="parent"
67.                app:layout_constraintTop_toTopOf="parent" />
68.            <TextView
70.                android:id="@+id/buju_gourment_count"
71.                android:layout_width="8dp"
72.                android:layout_height="wrap_content"
73.                android:layout_marginLeft="40dp"
74.                android:text="0"
75.                android:textColor="#60000000"
76.                app:layout_constraintBottom_toBottomOf="parent"
77.                app:layout_constraintStart_toStartOf="@id/buju_gourment_praise"
78.                app:layout_constraintTop_toTopOf="parent" />
79.        </android.support.constraint.ConstraintLayout>
80.    </LinearLayout>
```

第 7 行代码主要用于设置 ListView 组件中每个 item 的高度。在 item 的 layout 文件中，用 android:layout_height 设置 item 的高度，运行软件，会发现该高度设置无效，这是因为 ListView 每行的高度是由 inflater 填充布局中高度最大的控件决定的，而在 item 的 layout 文件中为 item 设定 minHeight 是设置 ListView 的每个子 Item 高度的较简洁及有效的方法。

（3）在 java 文件夹中新建一个 ViewHolder.java 类，为 BaseAdapter 优化做准备，具体代码如下。

```
public class ViewHolder {
    public TextView comment;
    public ImageView image;
    public TextView date;
    public TextView name;
    public ImageView praise;
    public TextView praise_count;
}
```

（4）在 java/adapter 文件夹中新建一个 GourmetBaseAdapter.java 类，用作 ListVeiw 数据适配的适配器，该类继承 BaseAdapter，同时重写 getCount()等方法，具体代码如下。

```
1.    public class GourmetBaseAdapter extends BaseAdapter {
2.        List<HashMap> data;
3.        Context mContext;
4.        ViewHolder viewHolder = null;
5.        public GourmetBaseAdapter(List<HashMap> mydata, Context myContext) {
6.            data = mydata;
7.            mContext = myContext;
8.        }
9.        public int getCount() {
```

```
10.            return data.size();
11.        }
12.        public Object getItem(int position) {
13.            return null;
14.        }
15.        public long getItemId(int position) {
16.            return 0;
17.        }
18.        public View getView(final int position, View convertView, ViewGroup parent) {
19.            if (convertView == null) {
20.                viewHolder = new ViewHolder();
21.                LayoutInflater mInflater = LayoutInflater.from(mContext);
22.                convertView = mInflater.inflate(R.layout.buju_gourmet_listview, null);
23.                viewHolder.name = (TextView) convertView
    .findViewById(R.id.buju_gourmet_username);
24.                viewHolder.date = (TextView) convertView
    .findViewById(R.id.buju_gourmet_time);
25.                viewHolder.comment = (TextView) convertView
    .findViewById(R.id.buju_gourmet_text);
26.                viewHolder.image = (ImageView) convertView
    .findViewById(R.id.buju_gourmet_pic);
27.                viewHolder.praise = (ImageView) convertView
    .findViewById(R.id.buju_gourmet_praise);
28.                viewHolder.praise_count = (TextView) convertView
    .findViewById(R.id.buju_gourmet_count);
29.                convertView.setTag(viewHolder);
30.            } else {
31.                viewHolder = (ViewHolder) convertView.getTag();
32.            }
33.            viewHolder.name.setText(data.get(position).get("name").toString());
34.            viewHolder.date.setText(data.get(position).get("date").toString());
35.            viewHolder.comment.setText(data.get(position).get("comment").toString());
36.            viewHolder.image.setBackgroundResource((Integer) data.get(position).get("image"));
37.            return convertView;
38.    }
```

第 5~8 行代码用于初始化数据。

第 10 行代码用于设置 ListView 需要显示的数据数量。

第 18 行代码中的 getView() 方法用于实现返回每一项的显示内容。

第 19 行代码的优化主要思路表现为若 convertView 为空则为其创建一个 View（一般在第一次加载时创建）；若 convertView 已加载过则直接使用缓存的数据，不需要重新加载。convertView 是刚刚离开屏幕的 View，可以复用。

第 21 行代码用于创建布局加载器。

第 22 行代码用于利用布局加载器将用于规范 ListView 每一项显示外观的布局加载到适配器中。由于只需要将 XML 转化为 View，并不涉及具体的布局，所以第二个参数通常设置为 null。

第 23~28 行代码用于为 viewHolder 的属性赋值。

第 29 行代码用于设置 convertView 的标签为 viewHolder，以便后面引用。

第 33～36 行代码用于设置各组件数据。

 小贴士

在 Android 中，ListView 的常用适配器通常使用 convertView 来缓存视图 View，以提高 ListView 的 item View 加载效率。建议在 Adapter 的 getView 中先判断 convertView 是否为空，如果非空，则直接对 convertView 复用；如果为空，创建新的 View。

（5）打开 Gourmet_fragment.java 程序，在该程序中添加相关代码以实现相关功能，具体代码如下。

```java
1.   public class Gourmet_fragment extends Fragment {
2.       private ListView mylistview;
3.       private int images[] = {
4.           R.drawable.p1_gourmet, R.drawable.p2_gourmet,
5.           R.drawable.p3_gourmet, R.drawable.p4_gourmet,
6.           R.drawable.p5_gourmet, R.drawable.p6_gourmet,
7.           R.drawable.p7_gourmet, R.drawable.p8_gourmet
8.       };//每一项分享的图片
9.       //每一项分享内容
10.      private String[] comment = { "减肥干吗  又不是吃不起", "这样的馒头 ，感觉能吃一筐", "给你一个爱上烘焙的理由", "不是我瘦不下来 是敌人太强", "一场咖啡与鲜花的比赛", "美食是灵魂伴侣", "吃货的幸福世界", "美得舍不得下咽" };
11.      private String[] date = { "2016 年 7 月", "2016 年 9 月", "2017 年 1 月", "2017 年 2 月", "2017 年 10 月", "2018 年 5 月", "2018 年 9 月", "2018 年 10 月" };//每一项分享时间
12.      private String[] name = { "叶德娴", "刘芸", "徐自贤 ", "丁志诚 ", "梁文道 ", "张笛 ", "杨若兮 ", "王丽达 " };//每一项的分享者
13.      private List<HashMap> dataList;
14.      GourmetBaseAdapter myBaseAdapter;//自定义适配器
15.      public View onCreateView(LayoutInflater inflater,@Nullable ViewGroup container,Bundle savedInstanceState) {
16.          //利用布局加载器加载个人中心布局，将其转换为 View
17.          View view = inflater.inflate(R.layout.frag_gourmet, null);
18.          initdata();
19.          myBaseAdapter = new GourmetBaseAdapter(dataList, this.getActivity());
20.          mylistview = (ListView) view.findViewById(R.id.gourment_frag_listView1);
21.          mylistview.setAdapter(myBaseAdapter);
22.          return view;
23.      }
24.      //初始化用于显示 listView 的数据
25.      private void initdata() {
26.          dataList = new ArrayList();
27.          for (int i = 0; i < date.length; i++) {
28.              HashMap hm = new HashMap();
29.              hm.put("image", images[i]);
30.              hm.put("date", date[i]);
31.              hm.put("comment", comment[i]);
32.              hm.put("name", name[i]);
33.              dataList.add(hm);
```

```
34.     }
35.   }
36. }
```

7.7 实现"吃货驾到"的点赞功能

1. 知识点

➤ BaseAdapter 中单击监听事件的方法。
➤ BaseAdapter 的 notifyDataSetInvalidated()方法的使用。

2. 工作任务

在"吃货驾到"页面中单击图 7-9 中标记 1 处的心形目标（点赞），能够将点赞数在标记 2 处进行显示。

图 7-9 点赞效果图

3. 操作流程

（1）在项目中打开 GourmetBaseAdapter.java 程序，并在该程序中定义一个 int count[]数组，此数组用于保存每一项的点赞数，并在结构化方法 GourmetBaseAdapter()中进行初始化，具体代码如下（第 4 行代码根据要显示的数据项数定义数组）。

```
1. public GourmetBaseAdapter(List<HashMap> mydata, Context myContext) {
2.     data = mydata;//要显示的数据
3.     mContext = myContext;//上下文
4.     count=new int[mydata.size()];
5. }
```

（2）在 getView( )方法中添加如下代码。

```
1. viewHolder.praise_count.setText(count[position]+"");
2. viewHolder.praise.setOnClickListener(new View.OnClickListener() {
3.     public void onClick(View v) {
4.         count[position]++;
5.         GourmetBaseAdapter.this.notifyDataSetChanged();
6.     }
7. });
```

第 1 行代码用于将 intcount[]数组保存的点赞数显示在 textView 上。
第 2～7 行代码用于实现单击心形图标则点赞数加 1，同时刷新界面。
第 5 行代码用于实现界面刷新功能。

小贴士

本工作任务虽然实现了点赞功能，但点赞的数据未保存，因此每次加载页面后，点赞数都会恢复为 0。

7.8 实现"吃货驾到"的功能菜单

1. 知识点

➢ 上下文菜单的添加方法。
➢ 上下文菜单选项功能的实现方法。
➢ 普通对话框的实现方法。

2. 工作任务

当长按"吃货驾到"页面中的某个条目时显示如图 7-10（a）所示的上下文菜单。当选择"微信分享"选项时弹出"确认"对话框（见图 7-10（b））；当单击该对话框中的"确定"按钮时，显示"微信分享成功"消息框；当单击该对话框中的"取消"按钮时，显示"你不微信分享了"消息框。当选择"收藏"选项时，显示"晚一点帮你收藏"消息框。当选择"删除"选项时，删除当前长按条目。

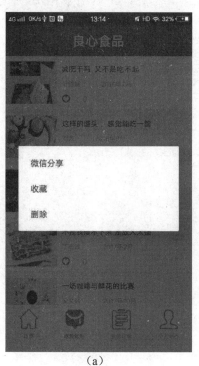

(a)　　　　　　　　　　　　(b)

图 7-10　"吃货驾到"功能菜单效果图

3. 操作流程

（1）打开 Gourmet_fragment.java 程序，并在该程序中重写 onCreateContextMenu()方法，创建上下文菜单（ContextMenu），具体代码如下（第 3~4 行代码用于添加菜单选项）。

```
1.  public void onCreateContextMenu(ContextMenu menu, View v, ContextMenuInfo menuInfo) {
2.      menu.add(0, 1, Menu.NONE, "微信分享");
3.      menu.add(0, 2, Menu.NONE, "收藏");
4.      menu.add(0, 3, Menu.NONE, "删除");
5.      super.onCreateContextMenu(menu, v, menuInfo);
6.  }
```

（2）在 onCreateView()方法中添加注册菜单方法 registerForContextMenu()，实现 ListView 组件与上下文菜单的绑定，具体代码如下。

```
this.registerForContextMenu(mylistview);
```

（3）创建 dialog()方法，用于创建普通对话框。当用户选择菜单中的"微信分享"选项时调用此对话框实现交互功能，具体代码如下。

```
1.  public void dialog(){
2.      AlertDialog.Builder mydialog=new AlertDialog.Builder(getActivity());
3.      mydialog.setTitle("确认");//设置对话框标题
4.      mydialog.setIcon(android.R.drawable.ic_dialog_alert); //设置对话框图标
5.      mydialog.setMessage("您确认要在微信分享");//设置对话框消息
6.      //为对话框设置一个"取消"按钮，并通过 new DialogInterface.OnClickListener()方法为按钮添加一个单击监听事件
7.      mydialog.setNegativeButton("取消",new DialogInterface.OnClickListener() {
8.          @SuppressLint("WrongConstant")
9.          public void onClick(DialogInterface dialogInterface,int i) {
10.             Toast.makeText(getActivity(),"你不微信分享了",2000).show();
11.         }
12.     });
13.     //为对话框设置一个"确定"按钮
14.     mydialog.setPositiveButton("确定",new DialogInterface.OnClickListener() {
15.         @SuppressLint("WrongConstant")
16.         public void onClick(DialogInterface dialogInterface,int i) {
17.             Toast.makeText(getActivity(),"微信分享成功",2000).show();
18.         }
19.     });
20.     mydialog.show();//显示对话框
21. }
```

（4）重写 onContextItemSelected()方法，实现菜单选择功能，具体代码如下。

```
1.  public boolean onContextItemSelected(MenuItem item) {
2.      final AdapterContextMenuInfo info=(AdapterContextMenuInfo) item.getMenuInfo();
3.      switch(item.getItemId()){
4.          case 1:
5.              Toast.makeText(getActivity(),"晚一点帮你分享",2000).show();
6.              break;
7.          case 2:
```

```
8.              Toast.makeText(getActivity(),"晚一点帮你收藏",2000).show();
9.              break;
10.         case 3:
11.             dataList.remove(info.position);//从当前 ListView 中移除当前选中的 item
12.             myBaseAdapter.notifyDataSetInvalidated();
13.             break;
14.     }
15.     return super.onContextItemSelected(item);
16. }
```

第 2 行代码用于获取 AdapterContextMenuInfo 对象，info 用于获取在 ListView 中选中被选项的序号，从而确定要删除的选项位置。

第 8 章 "我的订单"模块的设计

教学目标

◆ 了解 SQLite。
◆ 掌握创建数据库的方法。
◆ 掌握创建数据表的方法。
◆ 掌握查询数据表的方法。
◆ 掌握 SQLiteOpenHelper 的使用方法。

 8.1 工作任务概述

本章的主要工作任务是完成"我的订单"页面制作,需要完成以下工作子任务。
(1)完成"我的订单"页面的 UI 布局,效果如图 8-1(a)所示。
(2)制作"我的订单"页面的 Fragment。
(3)将"我的订单"页面添加到有底部导航的主 Activity 框架内(第 4 章创建的 MainActivity)。
(4)"吃货驾到"页面的"收藏"选项如图 8-1(b)所示。当选择了"吃货驾到"页面上下文菜单中的"收藏"选项时,将该分享内容的相关数据存储到本地 SQLite 数据库中,同时收藏信息在"我的订单"的"我的收藏"中显示出来,如图 8-1(c)所示。

(a)

(b)

(c)

图 8-1 "我的订单"UI 布局图

8.2 预备知识

1. SQLite 简介

SQLite 是一款轻型数据库,其功能是实现嵌入式,目前广泛应用于嵌入式产品。SQLite 占用的资源非常少,在嵌入式设备中,可能只需要几百 KB 的内存。

2. SQLite 的常用数据类型

SQLite 的常用数据类型如表 8-1 所示。

表 8-1 SQLite 的常用数据类型

类型名称	类 型	描 述
char	字符型	char 数据类型用来存储定长非统一编码型的数据
varchar	字符型	varchar 数据类型用来存储变长非统一编码型字符数据
integer	整型	integer 数据类型是一个带符号的整数,根据该整数值的大小存储在 1 字节、2 字节、3 字节、4 字节、6 字节或 8 字节中
real	近似数值型	real 数据类型与浮点数类似,是近似数值类型。它可以表示数值在-3.40E+38 到 3.40E+38 之间的浮点数
text	字符型	text 数据类型用来存储大量的非统一编码型字符数据。这种数据类型最多可以有 $2^{31}-1$ 个或 20 亿个字符
blob	blob 数据块	blob 数据类型按照输入的数据格式进行存储,不改变输入的数据格式
datetime	日期时间型	datetime 数据类型用来表示日期和时间。这种数据类型存储从 1753 年 1 月 1 日到 9999 年 12 月 31 日之间所有的日期和时间数据,精确到 1/300s 或 3.33ms

3. SQLite 编程简介

SQLite 的操作一般包括以下几种。

(1)创建和打开数据库。在 Android 中,使用 SQLiteDatabase 的静态方法 Context.openOrCreateDatabase(String name,int mode,CursorFactory factory)打开或创建一个数据库。其中,参数 String name 是数据库的保存路径及名称;参数 int mode 是操作数据库的模式;参数 CursorFactory factory 用于当打开的数据库执行查询语句时创建一个 Cursor 对象,这时会调用 Cursor 工厂类 factory,可以填写默认值 null。Context.openOrCreateDateBase()方法会自动检测是否存在要打开的数据库,存在则打开,不存在则创建一个数据库,创建成功则返回一个 SQLiteDatabase 对象,否则抛出异常 FileNotFoundException。

创建一个数据库 demo.db,实现代码如下。

```
SQLiteDatabase  db=SQLiteDatabase.openOrCreateDatabase("/data/data/com.demo.db/databases/demo.db",
Context.MODE_PRIVATE null);
```

(2)创建表。

第一步:编写创建表的 SQL 语句。

第二步：调用 SQLiteDatabase 的 execSQL()方法来执行 SQL 语句。

下面的代码创建了 stu_table 数据表，属性列为_id（主键并且自动增加）、sname（学生姓名）、snumber（学号）。

```
1.    private void createTable(SQLiteDatabase db){
2.        String stu_table="create table stu_table(_id integer primary key autoincrement,sname text,snumber text)";
3.        db.execSQL(stu_table);
4.    }
```

如果当前创建的数据表名已经存在，即与已经存在的表名、视图名和索引名冲突，那么本次创建操作将失败并报错，此时可以加上 if not exists 从句，那么本次创建操作将不会受到任何影响，即不会有错误抛出。

（3）向表中添加一条数据。插入数据有两种方法。

方法一：借助 SQLiteDatabase 的 insert(String table,String nullColumnHack,ContentValues values)方法。

- table：表名称。
- nullColumnHack：空列的默认值。
- ContentValues values：一个封装了列名称和列值的 ContentValues 类型的 Map。

方法一的代码如下。

```
1.    private void insert(SQLiteDatabase db){
2.        ContentValues cValue = new ContentValues();//实例化常量值
3.        cValue.put("sname","xiaoming");//添加用户名
4.        cValue.put("snumber","01005");//添加密码
5.        db.insert("stu_table",null,cValue); //调用 insert()方法插入数据
6.    }
```

方法二：编写插入数据的 SQL 语句，直接调用 SQLiteDatabase 的 execSQL()方法来执行该 SQL 语句。方法二的代码如下。

```
1.    private void insert(SQLiteDatabase db){
2.        //插入数据的 SQL 语句
3.        String stu_sql="insert into stu_table(sname,snumber) values('xiaoming','01005')";
4.        db.execSQL(sql); //执行 SQL 语句
5.    }
```

（4）查询表中的某条数据。

在 Android 中查询数据是通过 Cursor 类来实现的，使用 SQLiteDatabase.query()方可以得到一个 Cursor 对象，该 Cursor 对象指向每一条数据。查询表中的某条数据的具体方法如下。

```
public Cursor query(String table,String[]columns,String selection,String[]selectionArgs,String groupBy,String having,String orderBy,String limit);
```

- table：表名称，指定查询的表名。
- columns：列名称数组，指定查询的列名。
- selection：条件字句，相当于 where，指定 where 的约束条件。
- selectionArgs：条件字句，参数数组，为 where 中的点位符提供具体的值。
- groupBy：分组列，指定需要分组的列。

- having：分组条件，对分组后的结果进一步约束。
- orderBy：排序列，指定查询结果的排序方式。
- limit：分页查询限制。
- Cursor：返回值，相当于结果集 ResultSet。

Cursor 是一个游标接口，提供遍历查询结果的方法，如移动指针方法 move()、获得列值方法 getString()等，具体如表 8-2 所示。

表 8-2 Cursor 常用方法及其作用

方 法 名 称	作 用
getCount()	获得总的数据项数
isFirst()	判断是否为第一条记录
isLast()	判断是否为最后一条记录
moveToFirst()	移动到第一条记录
moveToLast()	移动到最后一条记录
move(int offset)	移动到指定记录
moveToNext()	移动到下一条记录
moveToPrevious()	移动到上一条记录
getColumnIndexOrThrow(String columnName)	根据列名称获得列索引
getInt(int columnIndex)	获得指定列索引的Int类型值
getString(int columnIndex)	获得指定列索引的String类型值

用 Cursor 来查询数据库中的数据，具体代码如下。

```
1.   private void query(SQLiteDatabase db) {
2.       //查询获得游标
3.       Cursor cursor = db.query ("usertable",null,null,null,null,null,null);
4.       //判断游标是否为空
5.       if(cursor.moveToFirst() {
6.           //遍历游标
7.           for(int i=0;i<cursor.getCount();i++){
8.               cursor.move(i);
9.               //获得 ID
10.              int id = cursor.getInt(0);
11.              //获得姓名
12.              String sname=cursor.getString(1);
13.              //获得学号
14.              String snumber=cursor.getString(2);
15.              //输出信息
16.      Log.("myinfo",+id+"+"+sname+":"+snumber)
17.          }
18.      }
19.  }
```

（5）从表中删除数据。

方法一：调用 SQLiteDatabase 的 delete(String table,String whereClause,String[]whereArgs) 方法。

- table：表名称。
- whereClause：删除条件。
- whereArgs：删除条件值数组。

方法一的代码如下。

```
1.   private void delete(SQLiteDatabase db) {
2.       String whereClause = "_id=?";//删除条件
3.       String[] whereArgs = {String.valueOf(2)}; //删除条件参数
4.       db.delete("stu_table",whereClause,whereArgs); //执行删除
5.   }
```

方法二：编写删除数据的 SQL 语句，调用 SQLiteDatabase 的 execSQL()方法执行该 SQL 语句。

方法二的代码如下。

```
1.   private void delete(SQLiteDatabase db) {
2.       //删除数据的 SQL 语句
3.       String sql = "delete from stu_table where _id = 6";
4.       //执行 SQL 语句
5.       db.execSQL(sql);
6.   }
```

（6）修改表中的数据。

方法一：调用 SQLiteDatabase 的 update(String table,ContentValues values,String whereClause, String[]whereArgs)方法。

- table：表名称。
- values：更新列 ContentValues 类型的键值对 Key-Value。
- whereClause：更新条件。
- whereArgs：更新条件值数组。

方法一的代码如下。

```
1.   private void update(SQLiteDatabase db) {
2.       //实例化内容值
3.       ContentValues values = new ContentValues();
4.       //在 values 中添加内容
5.       values.put("snumber","101003");
6.       //修改条件
7.       String whereClause = "_id=?";
8.       //修改条件参数
9.       String[] whereArgs={String.valuesOf(1)};
10.      //根据修改条件修改数据
11.      db.update("usertable",values,whereClause,whereArgs);
12.  }
```

方法二：编写更新数据的 SQL 语句，调用 SQLiteDatabase 的 execSQL()方法执行该 SQL 语句。

方法二的代码如下。

```
1.    private void update(SQLiteDatabase db){
2.        //更新数据的 SQL 语句
2.        String sql = "update stu_table set snumber = 654321 where _id = 1";
4.        //执行 SQL 语句
5.        db.execSQL(sql);
6.    }
```

（7）关闭数据库。关闭数据库很重要，但也是容易被忘记的。关闭数据库具体代码如下。

```
db.close();
```

（8）删除指定表。编写删除表的 SQL 语句，调用 SQLiteDatabase 的 execSQL()方法执行该 SQL 语句，具体代码如下。

```
1.    private void drop(SQLiteDatabase db){
2.        //删除表的 SQL 语句
2.        String sql ="DROP TABLE stu_table";
4.        //执行 SQL 语句
5.        db.execSQL(sql);
6.    }
```

（9）删除数据库。直接使用 deletDatabase()方法即可删除一个数据库，具体代码如下。

```
This. deletDatabase("demo.db");
```

4．SQLiteOpenHelper 类

SQLiteOpenHelper 类是 SQLiteDatabase 的一个辅助类，主要用于生成数据库，并对数据库的版本进行管理。

（1）构造方法：public ClassName(Context context,String name,CursorFactory factory,int version)。

- context：上下文对象（MainActivity.this）。
- name：数据库的名称。
- factory：创建 Cursor 对象，如果使用默认的 factory，则将该参数设置为 null。
- version：数据库的版本，版本号从 0 开始依次递增。

（2）两个回调函数。SQLiteOpenHelper 类是一个抽象类，在使用该类时通常需要继承它并实现里面的两个函数。

① onCreate(SQLiteDatabase)函数：一般在数据库第一次生成时，即创建数据库时会调用这个函数。一般在这个函数中生成数据表。

② onUpgrade(SQLiteDatabase,int,int)函数：当数据库需要升级时，Android 会主动调用这个函数。一般在这个函数中删除数据表并建立新的数据表，是否还需要进行其他操作完全取决于应用的需求。

代码如下。

```
1.    public class MyDatabaseOpenHelper extends SQLiteOpenHelper {
2.        private static final String db_name = "mydata.db"; //数据库名称
3.        private static final int version = 1; //数据库版本
```

```
4.     public DBOpenHelper(Context context, String db_name, SQLiteDatabase.CursorFactory factory,
       int version) {
5.         super(context, name, factory, version);
6.     }
7.     //该方法没有数据库存在才会执行
8.     public void onCreate(SQLiteDatabase db) {
9.         Log.i("Log","没有数据库,创建数据库");
10.        String sql_message = "create table t_message (id int primary key,userName varchar(50),
       lastMessage varchar(50),datetime   varchar(50))"; //创建表语句
11.        db.execSQL(sql_message); //执行创建表语句
12.    }
13.    //数据库更新才会执行
14.    public void onUpgrade(SQLiteDatabase db, int oldVersion, int newVersion) {
15.        Log.i("updateLog","数据库更新了！");
16.    }
17. }
```

（3）SQLiteDatabase 类与 SQLiteOpenHelper 类的完美配合。SQLiteOpenHelper 类通过调用 getWritableDatabase()方法或 getReadableDatabase()方法可以创建一个 SQLiteDatabase 类，具体代码如下。

```
1.  public class MainActivity extends Activity {
2.      protected void onCreate(Bundle savedInstanceState) {
3.          super.onCreate(savedInstanceState);
4.          setContentView(R.layout.activity_main);
5.          MyDatabaseOpenHelper helper = new MyDatabaseOpenHelper(MainActivity.this);
6.          SQLiteDatabase sqliteDatabase = helper.getWritableDatabase();
7.      }
8.  }
```

5. Android 获取图片资源的方式

（1）图片放在 sdcard 中。

```
Bitmap imageBitmap = BitmapFactory.decodeFile(path)
```

path 是图片的路径，根目录是/sdcard。

（2）图片放在项目的 res 文件夹中。

```
//得到 application 对象
ApplicationInfo appInfo = getApplicationInfo();
//得到该图片的 ID（name 是该图片的名称，drawable 是该图片存放的目录，appInfo. packageName 是应用程序的包）
int resID = getResources().getIdentifier(name,drawable,appInfo.packageName);
```

Android 获取图片的具体代码如下。

```
1.  public Bitmap getRes(String name)
2.  {
3.      ApplicationInfo appInfo = getApplicationInfo();
4.      int resID = getResources().getIdentifier(name, drawable, appInfo.packageName);
5.      return BitmapFactory.decodeResource(getResources(), resID);
6.  }
```

8.3 热身任务

本节的热身任务为"开心小秘书"。

1. 任务说明

"开心小秘书"用于管理日常信息,本项热身任务除了需要完成如图 8-2 所示的布局效果,还需要实现以下功能。

(a)　　　　　　　　　　　　　(b)

图 8-2 "开心小秘书"布局效果

(1) 创建一个本地 SQLite 数据库 sec。
(2) 在数据库 sec 中创建一个数据表 information,该表的结构如表 8-3 所示。

表 8-3 information 的结构

序号	字段名	类型	备注说明
1	_id	text	主键
2	title	text	标题
3	message	text	信息内容
4	date	text	日期

(3) 实现"添加"按钮功能:当单击"添加"按钮后将文本框中的信息添加至数据库中,若添加成功,则显示消息"添加成功";否则,显示消息"添加不成功"。

（4）实现"刷新"按钮功能：当单击"刷新"按钮后将数据表中的数据显示于按钮下方的 ListView 组件上。

（5）实现"查询"按钮功能：当在 id 值编辑框中输入 id 值后单击"查询"按钮，将数据表中与输入 id 值一致的各项数据显示在如图 8-2（b）所示的添加数据区域相对应的文本框中。

（6）实现"修改"按钮功能：当在如图 8-2（b）所示的添加数据区域修改数据时（如将 day 改成 today），单击"修改"按钮，将数据表_id 与 id 编辑框中值相同的数据按最新数据修改。

（7）实现"删除"按钮功能：当在 id 值编辑框中输入要删除的数据记录的 id 值后，再单击"删除"按钮，则将数据表_id 与 id 编辑框中值相同的数据记录删除。

2. 操作步骤

（1）创建一个 Android 项目。

（2）将图片复制到项目中的 drawable 文件夹中。

（3）打开 activity_main.xml 文件并完成布局效果，具体代码如下。

```xml
1.  <?xml version="1.0" encoding="utf-8"?>
2.  <LinearLayout xmlns:android="http://schemas.android.com/apk/res/android"
3.      xmlns:app="http://schemas.android.com/apk/res-auto"
4.      xmlns:tools="http://schemas.android.com/tools"
5.      android:layout_width="match_parent"
6.      android:layout_height="match_parent"
7.      android:orientation="vertical">
8.      <ImageView
9.          android:id="@+id/imageView"
10.         android:layout_width="match_parent"
11.         android:layout_height="wrap_content"
12.         android:src="@mipmap/sct" />
13.     <EditText
14.         android:id="@+id/editText2"
15.         android:layout_width="match_parent"
16.         android:layout_height="wrap_content"
17.         android:ems="10"
18.         android:hint="id 值不可以重复" />
19.     <EditText
20.         android:id="@+id/editText"
21.         android:layout_width="match_parent"
22.         android:layout_height="wrap_content"
23.         android:ems="10"
24.         android:hint="标题"
25.         android:inputType="textPersonName" />
26.     <EditText
27.         android:id="@+id/editText3"
28.         android:layout_width="match_parent"
29.         android:layout_height="wrap_content"
30.         android:ems="10"
31.         android:hint="描述"
32.         android:inputType="textMultiLine" />
```

```
33.    <EditText
34.        android:id="@+id/editText4"
35.        android:layout_width="match_parent"
36.        android:layout_height="wrap_content"
37.        android:ems="10"
38.        android:hint="日期"
39.        android:inputType="date" />
40.    <LinearLayout
41.        android:layout_width="match_parent"
42.        android:layout_height="wrap_content"
43.        android:orientation="horizontal">
44.        <Button
45.            android:id="@+id/button"
46.            android:layout_width="wrap_content"
47.            android:layout_height="wrap_content"
48.            android:layout_weight="1"
49.            android:background="#436EEE"
50.            android:onClick="add"
51.            android:text="添加"
52.            android:textColor="#ffffff" />
53.        <Button
54.            android:id="@+id/button2"
55.            android:layout_width="wrap_content"
56.            android:layout_height="wrap_content"
57.            android:layout_weight="1"
58.            android:background="#436EEE"
59.            android:onClick="refresh"
60.            android:text="刷新"
61.            android:textColor="#ffffff" />
62.        <Button
63.            android:id="@+id/button4"
64.            android:layout_width="wrap_content"
65.            android:layout_height="wrap_content"
66.            android:layout_weight="1"
67.            android:background="#436EEE"
68.            android:text="查询"
69.            android:onClick="find"
70.            android:textColor="#ffffff" />
71.        <Button
72.            android:id="@+id/button5"
73.            android:layout_width="wrap_content"
74.            android:layout_height="wrap_content"
75.            android:layout_weight="1"
76.            android:onClick="update"
77.            android:background="#436EEE"
78.            android:text="修改"
79.            android:textColor="#ffffff" />
80.        <Button
81.            android:id="@+id/button6"
```

```
82.        android:layout_width="wrap_content"
83.        android:layout_height="wrap_content"
84.        android:layout_weight="1"
85.        android:background="#436EEE"
86.        android:onClick="delete"
87.        android:text="删除"
88.        android:textColor="#ffffff" />
89.    </LinearLayout>
90.    <ListView
91.        android:id="@+id/ListView1"
92.        android:layout_width="match_parent"
93.        android:layout_height="wrap_content" />
94. </LinearLayout>
```

第 59 行代码的 onClick 属性设置了单击此组件所调用的方法，此方法要在 java 源程序中创建。

（4）在项目的\res\layout 文件夹中添加 item.xml 文件，该文件用于规范 ListView 组件每个列表选项的界面布局。item.xml 文件代码如下。

```
1.  <?xml version="1.0" encoding="utf-8"?>
2.  <LinearLayout xmlns:android="http://schemas.android.com/apk/res/android"
3.      android:layout_width="match_parent"
4.      android:layout_height="match_parent">
5.  //textView1 用于显示数据表中的 id 值
6.  <TextView
7.      android:id="@+id/textView1"
8.      android:layout_width="wrap_content"
9.      android:layout_height="wrap_content"
10.     android:layout_weight="1"
11.     android:text="TextView" />
12. //textView2 用于显示数据表中的 title 值
13. <TextView
14.     android:id="@+id/textView2"
15.     android:layout_width="wrap_content"
16.     android:layout_height="wrap_content"
17.     android:layout_weight="1"
18.     android:text="TextView" />
19. //textView3 用于显示数据表中的 message 值
20. <TextView
21.     android:id="@+id/textView3"
22.     android:layout_width="wrap_content"
23.     android:layout_height="wrap_content"
24.     android:layout_weight="1"
25.     android:text="TextView" />
26. //textView4 用于显示数据表中的 date 值
27. <TextView
28.     android:id="@+id/textView4"
29.     android:layout_width="wrap_content"
30.     android:layout_height="wrap_content"
```

```
31.            android:layout_weight="1"
32.            android:text="TextView" />
33. </LinearLayout>
```

（5）打开 java 文件夹中的 MainActivity.java 源程序，修改代码，实现相关功能，具体代码如下。

```
1.  public class MainActivity extends Activity {
2.      private EditText Et_id, Et_title, Et_message, Et_date;
3.      private SQLiteDatabase myDateBase;
4.      private ListView lv;
5.      protected void onCreate(Bundle savedInstanceState) {
6.          super.onCreate(savedInstanceState);
7.          setContentView(R.layout.activity_main);
8.          init();
9.          //创建或打开 SQLite 数据库 sec
10.         myDateBase = this.openOrCreateDatabase("sec", Context.MODE_PRIVATE, null);
11.         createTabe();
12.     }
13.     //此方法用于创建数据表
14.     public void createTabe() {
15.         String cmd = "create table if not exists information(_id text primary key ,title text,message text,date text)";//创建数据表的 SQL 语句
16.         myDateBase.execSQL(cmd);//执行 SQL 语句
17.     }
18.     //此方法用于初始化
19.     public void init() {
20.         Et_id = this.findViewById(R.id.editText2);
21.         Et_title = this.findViewById(R.id.editText);
22.         Et_message = this.findViewById(R.id.editText3);
23.         Et_date = this.findViewById(R.id.editText4);
24.         lv = this.findViewById(R.id.ListView1);
25.     }
26.     //此方法用于添加数据记录
27.     public void add(View v) {
28.         String id = Et_id.getText().toString();
29.         String title = Et_title.getText().toString();
30.         String message = Et_message.getText().toString();
31.         String date = Et_date.getText().toString();
32.         //创建 ContentValues，用于存放每条数据记录的字段值
33.         ContentValues value = new ContentValues();
34.         value.put("title", title);
35.         value.put("message", message);
36.         value.put("_id", id);
37.         value.put("date", date);
38.         //调用 insert()方法插入记录，插入数据成功则返回插入记录在数据表中的行号，插入数据不成功则返回数据值-1，可以依此来判断是否插入记录成功
39.         long i = myDateBase.insert("information", null, value);
40.         if (i >= 0) {
41.             Toast.makeText(this, "添加成功", Toast.LENGTH_SHORT).show();
```

```
42.            } else {
43.                Toast.makeText(this, "添加不成功", Toast.LENGTH_SHORT).show();
44.            }
45.        }
46.        //此方法用于将 information 数据表中的数据信息显示在 listView 组件上
47.        public void refresh(View v) {
48.            Cursor c = myDateBase.query("information", null, null, null, null, null, null);
49.            SimpleCursorAdapter adapter;
50.            adapter = new SimpleCursorAdapter(this, R.layout.item, c, new String[]{"_id", "title", "date", "message"}, new int[]{R.id.textView1, R.id.textView2, R.id.textView3, R.id.textView4});
51.            lv.setAdapter(adapter);
52.        }
53.        //此方法用于删除数据
54.        public void delete(View v) {
55.            String id = Et_id.getText().toString();
56.            //利用 delete()方法删除 id 值与 id 编辑框中与 id 值一致的数据记录,若删除成功则返回受影响的行号,否则返回 0。可用返回值来判断是否删除成功
57.            int i =myDateBase.delete("information", "_id=?", new String[]{id});
58.            if (i > 0) {
59.                Toast.makeText(this, "删除成功", Toast.LENGTH_SHORT).show();
60.            } else {
61.                Toast.makeText(this, "删除不成功", Toast.LENGTH_SHORT).show();
62.            }
63.        }
64.        //此方法用于根据文本编辑框的值修改数据表中该_id 值的数据记录信息
65.        public void update(View v) {
66.            String id = Et_id.getText().toString();
67.            String title = Et_title.getText().toString();
68.            String message = Et_message.getText().toString();
69.            String date = Et_date.getText().toString();
70.            ContentValues value = new ContentValues();
71.            value.put("title", title);
72.            value.put("message", message);
73.            value.put("date", date);
74.            myDateBase.update("information", value, "_id=?", new String[]{id});
75.        }
76.        //此方法用于在数据表中查找_id 值与编辑框中的 id 值一致的数据记录,并显示在页面上
77.        public void find(View v) {
78.            String value = Et_id.getText().toString();
79.            Cursor c = myDateBase.query("information", null, "_id=?", new String[]{value}, null, null, null);
80.            if(c.getCount() > 0) {//通过游标的记录数判断是否有满足条件的数据记录
81.                //将游标移动至第一条记录,由于此处的 id 是主键,因此查询结果只有一条记录
82.                c.moveToFirst();
83.                String id = c.getString(c.getColumnIndex("_id"));//获取游标中字段为_id 的数据记录
84.                Et_id.setText(id);
85.                String title = c.getString(c.getColumnIndex("title"));
86.                Et_title.setText(title);
87.                String message = c.getString(c.getColumnIndex("message"));
88.                Et_message.setText(message);
```

```
89.                        String date = c.getString(c.getColumnIndex("date"));
90.                        Et_date.setText(date);
91.            }
92.        }
93. }
```

第 15 行代码中的 if not exists 用于判断创建的表是否存在，因为在创建表的过程中若此表已存在，重复创建则会报错，所以可先通过 if not exists 语句进行判断，当该表不存在时才创建表。

第 27 行代码中的参数 View v 的作用是在 XML 文件的组件中通过添加 onClick 属性调用 add() 方法。此参数不可少。

第 32～37 行代码创建了一个 ContentValues，用于插入数据记录。而这里必须用 ContentValues 的原因是 SQLiteDataBase 对象的 insert() 接口：public long insert(String table,String nullColumnHack, ContentValues values)。同理，在进行 SQLiteDataBase 对象的数据更新时也要将数据转换成一个 ContentValues。

👉 小贴士

在进行 SQLite 数据库操作时，为了更清晰地了解数据库的情况，通常需要借助 adb.exe 工具进入 adb 内核进行数据库的相关操作，具体做法如下。

第一步：进入 adb 内核。在 Sdk 目录下找到 platform-tools 文件夹，adb.exe 就在这个文件夹下。在文件夹空白区域按住 Shift 键，同时单击鼠标右键，在弹出的快捷菜单中选择"在此处打开命令窗口"，会弹出如图 8-3（a）所示的 cmd 窗口（也可以直接打开 cmd，然后进入相应的路径），在 cmd 窗口中输入 adb shell，命令执行后如图 8-3（b）所示。

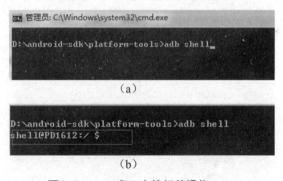

图 8-3　cmd 窗口中的相关操作

第二步：进入应用数据库目录，具体操作如下（com.xxx.xxx 是项目包名）。

cd /data/data/com.xxx.xxx/databases

第三步：验证数据库文件。

ls

第四步：使用 sqlite3 命令打开数据库文件（xxx.db 是数据库名）。

sqlite3 xxx.db

第五步：查看表和创建表语句。

sqlite> .tables

 8.4　创建"我的订单"Fragment

1. 知识点

➢ ListView 的添加方法。
➢ ListView 的常用属性。
➢ Fragment 创建方法。

2. 工作任务

制作"我的订单"模块的 UI 布局并创建相应的 Fragment。

3. 操作流程

（1）依次单击"File"→"New"→"XML"→"Layout XML File"选项，创建 frag_order.xm 文件。

（2）在 frag_order.xml 文件中添加组件，实现如图 8-4 所示的布局效果。"我的订单"的 Component Tree 如图 8-5 所示。

图 8-4　"我的订单"布局效果

图 8-5 "我的订单"的 Component Tree

（3）打开 frag_order.xml 文件，修改各组件的相关属性，具体代码如下。

```xml
1.  <?xml version="1.0" encoding="utf-8"?>
2.  <LinearLayout xmlns:android="http://schemas.android.com/apk/res/android"
3.      android:layout_width="match_parent"
4.      android:layout_height="match_parent"
5.      android:orientation="vertical">
6.      <TextView
7.          android:id="@+id/frag_order_textView1"
8.          android:layout_width="wrap_content"
9.          android:layout_height="wrap_content"
10.         android:layout_marginTop="10dp"
11.         android:text="最近订单" />
12.     <ListView
13.         android:id="@+id/frag_order_orderList"
14.         android:layout_width="match_parent"
15.         android:layout_height="0dp"
16.         android:layout_weight="4">
17.     </ListView>
18.     <LinearLayout
19.         android:layout_width="match_parent"
20.         android:layout_height="0dp"
21.         android:layout_marginTop="10dp"
22.         android:layout_weight="5"
23.         android:orientation="vertical">
24.         <TextView
25.             android:id="@+id/frag_order_textView2"
26.             android:layout_width="wrap_content"
27.             android:layout_height="wrap_content"
28.             android:text="我的收藏" />
29.         <ListView
30.             android:id="@+id/frag_order_collectionList"
31.             android:layout_width="match_parent"
32.             android:layout_height="match_parent">
33.         </ListView>
34.     </LinearLayout>
35. </LinearLayout>
```

（4）在项目的 java\fragment 文件夹中新建 Order_Fragment.java，继承 Fragment，同时添加 onCreateView()方法。

（5）重写 Order_Fragment.java 中的 onCreateView()方法，具体代码及相关功能说明如下。

```
1.  public View onCreateView(LayoutInflater inflater, ViewGroup container, Bundle savedInstanceState) {
2.      // TODO Auto-generated method stub
3.      //利用布局加载器加载"我的订单"布局，将其转换为 View
4.      View view = inflater.inflate(R.layout.frag_order, null);
5.      return view; //返回 View
6.  }
```

8.5 将"我的订单"碎片组装至 App 主框架

1. 知识点

Fragment 动态加载方法。

2. 工作任务

将创建完成的"我的订单"（此时"我的订单"页面是空白的）碎片组装至"良心食品" App 主框架中，如图 8-6 所示。组装完成后单击 App"底部导航"栏中的"我的订单"（图 8-6 中标记 1 处）时，能够将"我的订单"碎片在 App 内（图 8-6 中标记 2 处）进行显示。

图 8-6 "我的订单"碎片组装效果图

3. 操作流程

在 project 视图中打开项目中的 MainActivity.java 源程序，修改 navigaion()方法，用于当切换至导航中的"我的订单"时将"我的订单"页面加载至程序主框架中，如图 8-7 所示。在原有程序代码的基础上添加以下代码。

```
transaction.replace(R.id.main_framelayout,new Order_Fragment());
```

图 8-7　navigation()方法代码

8.6　实现"最近订单"的数据显示

1. 知识点

➢ ListView 数据适配方法。
➢ SimpleAdapter 的使用方法。

2. 工作任务

将"最近订单"的数据显示在"我的订单"页面内。"最近订单"效果图如图 8-8 所示。

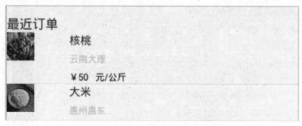

图 8-8　"最近订单"效果图

3. 操作流程

（1）在项目中的\res\layout 文件夹中添加 buju_order_orderlist.xml 文件，用于规范 ListView 选项元素的界面布局。"最近订单"的 Component Tree 如图 8-9 所示。

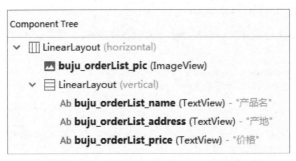

图 8-9　"最近订单"的 Component Tree

buju_order_orderlist.xml 文件代码如下。

```xml
1.  <?xml version="1.0" encoding="utf-8"?>
2.  <LinearLayout xmlns:android="http://schemas.android.com/apk/res/android"
3.      android:layout_width="match_parent"
4.      android:layout_height="match_parent"
5.      android:orientation="horizontal" >
6.      <ImageView
7.          android:id="@+id/buju_orderList_pic"
8.          android:layout_width="50dp"
9.          android:layout_height="30dp"
10.         android:layout_marginTop="10dp"
11.         android:paddingLeft="30dp"
12.         android:scaleType="fitStart"/>
13.     <LinearLayout
14.         android:layout_width="match_parent"
15.         android:layout_height="match_parent"
16.         android:layout_marginTop="5dp"
17.         android:layout_marginLeft="10dp"
18.         android:orientation="vertical" >
19.         <TextView
20.             android:id="@+id/buju_orderList_name"
21.             android:layout_width="wrap_content"
22.             android:layout_height="wrap_content"
23.             android:textSize="12sp"
24.             android:text="产品名"/>
25.         <TextView
26.             android:id="@+id/buju_orderList_address"
27.             android:layout_width="wrap_content"
28.             android:layout_height="wrap_content"
29.             android:layout_marginTop="5dp"
30.             android:textSize="10sp"
31.             android:textColor="#D1D1D1"
32.             android:text="产地" />
```

```
33.            <TextView
34.                android:id="@+id/buju_orderList_price"
35.                android:layout_width="wrap_content"
36.                android:layout_height="wrap_content"
37.                android:layout_marginTop="5dp"
38.                android:textColor="#D1D1D1"
39.                android:textSize="10sp"
40.                android:text="价格" />
41.        </LinearLayout>
42.    </LinearLayout>
```

（2）打开 Order_Fragment.java 源程序，在该程序中添加相应代码，实现显示"最近订单"数据的功能，具体代码如下。

```
1.  public class Order_Fragment extends Fragment {
2.      private int images[] = {R.drawable.po1_hetao,R.drawable.po2_mi,
    R.drawable.po3_jidan,R.drawable.po4_huasheng, R.drawable.po5_huajiao};
        private String[] price = {"￥50 元/公斤 ","￥10 元/公斤 ","￥25 元/公斤 ","￥30 元/公斤 ",
    "￥100 元/公斤 "};
3.      private String[] address = {"云南大理","惠州惠东","惠州农门","河源和平","四川汶川"};
4.      private String[] name = {"核桃","大米","鸡蛋","花生 ","花椒"};
5.      ArrayList dataList;
6.      ListView order;
7.      public View onCreateView(LayoutInflater inflater,@Nullable ViewGroup container,Bundle savedInstanceState) {
8.          //利用布局加载器加载"我的订单"的布局，将其转换为 View
9.          View view = inflater.inflate(R.layout.frag_order,null);
10.         init(view);
11.         showOrder();
12.         return view; //返回 View
13.     }
14. //初始化组件
15.     public void init(View view) {
16.         order = view.findViewById(R.id.frag_order_orderList);
17.     }
18. //将数据适配至 ListView 组件
19.     public void showOrder() {
20.         initData();
21.         //创建简单适配器：第一个参数表示上下文，第二个参数表示数据源，第三个参数显示
    数据的布局，第四个参数表示数据源哪些数据显示于布局，第五个参数表示数据对应显示至布局
    中的哪个组件。
22.         SimpleAdapter adapter = new SimpleAdapter(this.getActivity(),
    dataList,   R.layout.buju_order_orderlist, new String[]{"image","price","address","name"},
    new int[]{R.id.buju_orderList_pic,R.id.buju_orderList_price,R.id.buju_orderList_address,R.id.buju_orderList_name});
23.         order.setAdapter(adapter);
24.     }
25. //初始化"最近订单"数据
26.     public void initData() {
27.         dataList = new ArrayList();
28.         for (int i = 0; i < images.length; i++) {
```

```
29.            HashMap hm = new HashMap();
30.            hm.put("image",images[i]);
31.            hm.put("price",price[i]);
32.            hm.put("address",address[i]);
33.            hm.put("name",name[i]);
34.            dataList.add(hm);
35.        }
36.    }
37. }
```

 小贴士

利用适配器适配数据的大致流程：准备数据源（由第 27~33 行代码完成）→创建适配器（由第 21 行代码完成）→组件绑定适配器（由第 22 行代码完成）。

 8.7 实现"吃货驾到"的收藏功能

1. 知识点

➢ SQLiteOpenHelper 的使用方法。
➢ SQLiteDatabase 添加数据的方法。

2. 工作任务

本工作任务主要实现"吃货驾到"分享内容的上下文菜单中的"收藏"功能，如图 8-10 所示。当选择"收藏"选项时，将该分享内容的相关数据存储到本地数据库 collection.db 中，收藏成功后如图 8-11 所示。

图8-10 "吃货驾到"收藏功能

图8-11 收藏成功

3. 操作流程

（1）创建 DBOpenHelper.java 程序并继承 SQLiteOpenHelper，用于对 SQLite 数据库的管理，具体代码如下。

```
1.  public class DBOpenHelper extends SQLiteOpenHelper
2.  //构造函数4个参数的意义：参数 Context 为上下文对象；参数 name 为数据库的名称；参数 factory
    用于创建 Cursor 对象，该参数默认设置为 null；参数 vision 为数据库的版本
    //
3.      public DBOpenHelper(Context context, String name, SQLiteDatabase.CursorFactory factory, int version) {
4.          super(context, name, factory, version);
5.      }
6.  //在数据库第一次生成，即创建时会调用 onCreate()方法，一般在这个方法中生成数据库表
7.      public void onCreate(SQLiteDatabase sqLiteDatabase) {
8.          String create_table="create table if not exists collection_imf(_id integer primary key autoincrement,name text,date text,comment text,image BLOB)";
9.          sqLiteDatabase.execSQL(create_table);
10.     }
11. //当数据库需要升级时，Android 会主动调用 onUpgrade()方法。一般在这个方法中
    删除数据表，并建立新的数据表，是否需要进行其他操作完全取决于应用的需求
12.     public void onUpgrade(SQLiteDatabase sqLiteDatabase, int i, int i1) {
13.     }
14. }
```

（2）打开项目中的 Gourmet_fragment.java 程序，在该程序中添加 getPicture()方法，用于将 Drawalbe 图片转换成字节数组，以便于存储在数据库中，具体代码及相关说明如下。

```
1.  private byte[] getPicture(Drawable drawable) {
2.      if (drawable == null) {
3.          return null;
4.      }
5.      BitmapDrawable bd = (BitmapDrawable) drawable;//将 Drawable 转换成 BitmapDrawable 类型
6.      Bitmap bitmap = bd.getBitmap();//将 Drawable 转换成 Bitmap 类型
7.      ByteArrayOutputStream os = new ByteArrayOutputStream();//创建字节数组输出流
8.  //利用 Bitmap.compress()方法将图片压缩至数据流中
9.      bitmap.compress(Bitmap.CompressFormat.PNG, 100, os);
10.     return os.toByteArray();
11. }
```

（3）在 Gourmet_fragment.java 程序中添加 collection()方法，用于将收藏数据存储在 collection.db 中，具体代码及相关说明如下。

```
1.  public void collection(int select_index) {
2.  //打开或创建 collection.db 数据库
3.      openHelper = new DBOpenHelper(this.getActivity(), "collection.db",null, 1);
4.      SQLiteDatabase sqliteDatabase = openHelper.getWritableDatabase();//得到 SQLiteDatabase 数据库
5.      ContentValues values = new ContentValues();
6.      values.put("name", dataList.get(select_index).get("name").toString());
7.      values.put("date", dataList.get(select_index).get("date").toString());
```

```
8.      values.put("comment", dataList.get(select_index).get("comment") toString());
9.      values.put("image", getPicture(this.getResources().getDrawable((Integer) dataList.get(select_index).get("image"))));
10.      long i = sqliteDatabase.insert("collection_imf", null, values);
11.      if (i > -1) {
12.         oast.makeText(getActivity(), "亲！已收藏", Toast.LENGTH_SHORT).show();
13.      }
14.      sqliteDatabase.close();
15.  }
```

第 5～9 行代码创建一个 ContentValues，将要添加至数据库中的数据添加进去，为第 10 行代码将数据插入数据库中做准备。其中 select_index 记录了 ListView 中被选中的添加收藏的行序号。

（4）在 Gourmet_fragment.java 程序中的 onContextItemSelected()方法中调用 collection()方法。调用 collection()方法的程序段如图 8-12 所示。

```
public boolean onContextItemSelected(MenuItem item) {
    final AdapterView.AdapterContextMenuInfo info=(AdapterView.AdapterContextMenuInfo) item.getMenuInfo
    switch(item.getItemId()){
        case 1:
            dialog();
            break;
        case 2:
            collection(info.position);  //通过info.position获取当前项在ListView中的位置
            break;
        case 3:
            dataList.remove(info.position);
            myBaseAdapter.notifyDataSetInvalidated();
            break;
    }
    return super.onContextItemSelected(item);
}
```

图 8-12　调用 collection()方法的程序段

8.8　实现"我的订单"中"我的收藏"区域数据的显示

1. 知识点

➢ 用 SQLiteDatabase 查询数据的方法。
➢ BaseApater 的使用方法。

2. 工作任务

将在"吃货驾到"页面收藏的信息显示在"我的订单"的"我的收藏"中，实现如图 8-13 所示的效果。

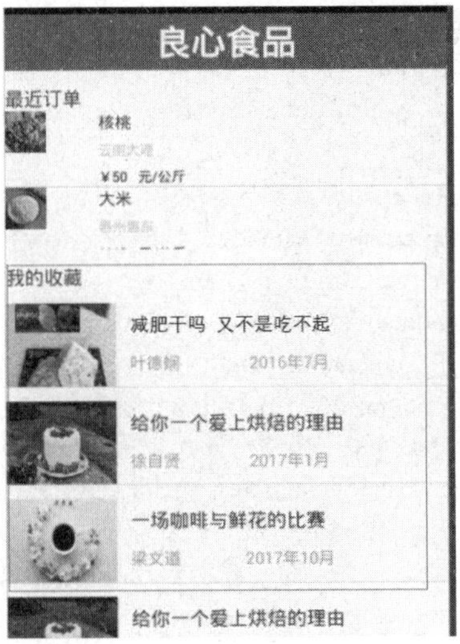

图 8-13 "我的收藏"效果图

3. 操作流程

（1）在 java/adapter 文件夹中新建一个 OrderBaseAdapter.java 类继承 BaseAdapter，该类用于数据的适配，具体代码及相关说明如下。

```
1.  public class OrderBaseAdapter extends BaseAdapter {
2.      List<HashMap<String,Object>> data;
3.      Context mContext;
4.      ViewHolder viewHolder = null;
5.      public OrderBaseAdapter(List<HashMap<String,Object>> mydata, Context myContext) {
6.          data = mydata;//要显示的数据
7.          mContext = myContext;//上下文
8.      }
9.      public int getCount() {
10.         // TODO Auto-generated method stub
11.         return data.size();
12.     }
13.     @Override
14.     public Object getItem(int position) {
15.         // TODO Auto-generated method stub
16.         return null;
17.     }
18.     @Override
19.     public long getItemId(int position) {
20.         // TODO Auto-generated method stub
21.         return 0;
22.     }
23.     @Override
24.     public View getView(int position, View convertView, ViewGroup parent) {
25.         if (convertView == null) {
26.             viewHolder = new ViewHolder();
27.             LayoutInflater mInflater = LayoutInflater.from(mContext);//产生布局加载器
```

```
28.                //利用布局加载器将用于规范 ListView 每一项显示外观的布局加载进适配器中,
                   由于只需要将 XML 转化为 View,并不涉及具体的布局,所以第二个参数通常设置为 null
29.                convertView = mInflater.inflate(R.layout.buju_order_collectionlist,
null);
30.                //对 viewHolder 的属性进行赋值
31.                viewHolder.name = (TextView)   convertView
findViewById(R.id.buju_order_collection_username);
32.                viewHolder.date = (TextView) convertView
findViewById(R.id.buju_order_collection_time);
33.                viewHolder.comment = (TextView) convertView
findViewById(R.id.buju_order_collection_text);
34.                viewHolder.image = (ImageView) convertView
findViewById(R.id.buju_order_collection_pic);
35.                //将 convertView 的标签设置为 viewHolder,便于后面引用
36.                convertView.setTag(viewHolder);
37.            } else {
38.                viewHolder = (ViewHolder) convertView.getTag();
39.            }
40.            viewHolder.name.setText(data.get(position).get("name").toString());
41.            viewHolder.date.setText(data.get(position).get("date").toString());
42.            viewHolder.comment.setText(data.get(position).get("comment").toString());
43.            Drawable bm=(Drawable) data.get(position).get("image");
44.            viewHolder.image.setBackground(bm);
45.            return convertView;
46.        }
47. }
```

(2) 打开 Order_Fragment.java 程序,在该程序中添加 getCollection()方法,用于从数据库中读取数据,具体代码及相关说明如下。

```
1.  private void getCollection() {
2.      CollectionList = new ArrayList();
3.      openHelper = new DBOpenHelper(this.getActivity(), "collection.db", null, 1);
4.      SQLiteDatabase sqliteDatabase = openHelper.getReadableDatabase();
5.      Cursor cursor = sqliteDatabase.query("collection_imf", null, null, null, null, null, null);
6.      //遍历数据
7.      if (cursor != null && cursor.getCount() != 0) {
8.          while (cursor.moveToNext()) {//获取数据
9.              HashMap hm = new HashMap();
10.             String name = cursor.getString(cursor.getColumnIndex("name"));
11.             String date = cursor.getString(cursor.getColumnIndex("date"));
12.             String comment = cursor.getString(cursor.getColumnIndex("comment"));
13.             byte[] b = cursor.getBlob(cursor.getColumnIndex("image"));
14.             //将获取的数据转换成 Drawable
15.             Bitmap bitmap = BitmapFactory.decodeByteArray(b, 0, b.length, null);
16.             Drawable drawable = new BitmapDrawable(bitmap);
17.             hm.put("name", name);
18.             hm.put("date", date);
19.             hm.put("comment", comment);
20.             hm.put("image", drawable);
```

```
21.            CollectionList.add(hm);
22.        }
23.    }
24. }
```

第 8~20 行代码利用 Cursor 将数据从数据库中逐条读取出来之后存放于 HashMap 中，并将所有的数据放入 CollectionList 中。

（3）在 Order_Fragment.java 程序中添加 showCollection()方法，用于将数据适配于"我的收藏"的 ListView 组件上，具体代码如下。

```
1. public void showCollection() {
2.     getCollection();
3.     OrderBaseAdapter myadapter=new OrderBaseAdapter(CollectionList,this.getActivity());
4.     collection.setAdapter(myadapter);
5. }
```

（4）修改 Order_Fragment.java 程序中的 init()方法，在该方法中添加如下代码。

```
collection = (ListView) view.findViewById(R.id.frag_order_collectionList);
```

添加上述代码后的 init()方法代码如图 8-14 所示。

```
//-------------初始化组件---------
public void init(View view) {
    order = view.findViewById(R.id.frag_order_orderList);
    collection = (ListView) view.findViewById(R.id.frag_order_collectionList);
}
```

图 8-14　添加上述代码后的 init()方法代码

（5）修改 Order_Fragment.java 程序中的 onCreateView()方法，调用 showCollection()方法，实现在"我的订单"Fragment 加载时显示"我的收藏"页面内容。修改后的 onCreateView()方法代码如图 8-15 所示。

```
public View onCreateView(LayoutInflater inflater,@Nullable ViewGroup container,Bundle savedIn
    View view = inflater.inflate(R.layout.frag_order, root: null);    //利用布局加载器载我的订
    init(view);
    showOrder();
    showCollection();
    return view;    //返回view
}
```

图 8-15　修改后的 onCreateView()方法代码

 第 9 章 登录验证

 教学目标

◆ 了解 Android 网络开发常用技术。
◆ 掌握利用 HttpURLConnection 类进行 Android 网络开发的方法。
◆ 掌握使用 SharedPreferences 对象存储数据的方法。
◆ 掌握利用 Android 原生技术解析 JSON 的方法。
◆ 掌握 ProgressDialog 的使用方法。

 9.1 工作任务概述

本章的主要工作任务是实现"登录验证"功能，需要完成以下工作子任务。

（1）实现在登录界面（见图 9-1（a））输入用户名及密码，单击"登入"按钮后，将用户名及用户密码上传至服务器，服务器对用户身份进行验证后，将验证结果返回客户端的功能。

（2）将服务器返回的信息进行 JSON 解析，判断用户个人信息是否通过验证，若通过验证则将用户名保存至 SharedPreferences 中，关闭登录界面，将"个人中心"的"待君登入"修改为服务器发回的用户名（见图 9-1（b））。

（a）

（b）

图 9-1 "登录验证"效果图

9.2 预备知识

9.2.1 SharedPreferences

1. SharedPreferences 概述

SharedPreferences 存储方式是一种 Android 中存储轻量级数据的方式，主要用来存储一些简单的配置信息，内部以 Map 方式进行存储，因此需要使用键值对提交和保存数据，保存的数据以 XML 格式存放在本地的/data/data/<packagename>/shares_prefs 文件夹下。SharedPreferences 有如下 3 个特点。

（1）使用简单，便于存储轻量级的数据。
（2）只支持 Java 基本数据类型，不支持自定义数据类型。
（3）属于单例对象，在整个应用内共享数据，无法在其他应用内共享数据。

2. 使用 SharedPreferences 对象存储数据

第一步：获得 SharedPreferences 对象。SharedPreferences 对象必须使用上下文获得。在使用 SharedPreferences 时注意先要获得上下文。获得 SharedPreferences 对象的方法如下。

```
SharedPreferences sharedPreferences = getSharedPreferences(参数一,参数二);
```

参数一：要保存的 XML 文件名，不同的文件名产生的对象不同，但同一文件名可以产生多个引用，从而可以保证数据共享。注意此处在指定参数一时，不用手动添加 xml 后缀，而由系统自动添加。

参数二：创建模式，常用的有 4 个值：MODE_PRIVATE、MODE_WORLD_READABLE、MODE_WORLD_WRITEABLE、MODE_APPEND。

第 1 个值使 SharedPreferences 存储的数据只能在本应用内获得；第 2 个值和第 3 个值分别使其他应用可以读、写本应用 SharedPreferences 存储的数据，由此可能带来安全问题；第 4 个值使系统先检查文件是否存在，若存在则在文件中追加内容，否则创建新文件。

第二步：获得 editor 对象。使用以上获得的 SharedPreferences 对象获得 editor 对象，方法如下。

```
Editor editor = sharedPreferences.edit();
```

第三步：对数据实现增、删、查改。
添加、修改数据：

```
editor.putString(key,value);
```

若此方法操作的键值对中的 key 不存在，则实现添加数据的功能；反之，则实现修改数据的功能。putString()方法用于添加字符串类型的 value，若要添加其他类型的 value，则需要替换 String。例如，若要添加 float 类型的 value，则需要使用 putFloat(key, value)方法。

删除数据：

```
editor.remove(key);
```

删除参数部分键的键值对。
清空数据：

```
editor.clear();
```

提交数据：

```
editor.commit;
```

第四步，查询数据。

```
String result = sharedPreferences.getString(key1,key2);
```

key1 是要查询的键，返回对应的值，当键不存在时，返回 key2 作为结果。

使用 SharedPreferences 对象存储数据的示例代码如下。

```
1.  protected void onCreate(Bundle savedInstanceState) {
2.      super.onCreate(savedInstanceState);
3.      setContentView(R.layout.activity_main);
4.      sharedPreferences = getSharedPreferences("info2", MODE_PRIVATE);
5.      editor = sharedPreferences.edit();
6.      et_key = (EditText) findViewById(R.id.et_key);
7.      et_value = (EditText) findViewById(R.id.et_value);
8.      et_query = (EditText) findViewById(R.id.et_query);
9.      tv_query = (TextView) findViewById(R.id.tv_content);
10.     et_delete = (EditText) findViewById(R.id.et_delete);
11.  public void insert(View v) {
12.     //获取键值数据
13.     String key = et_key.getText().toString().trim();
14.     String value = et_value.getText().toString().trim();
15.         //使用 editor 保存数据
16.     editor.putString(key, value);
17.     //注意一定要提交数据，此步骤非常容易被忽略
18.     editor.commit();
19.  }
20.  public void query(View v) {
21.         //获得查询的键
22.         String query_text = et_query.getText().toString().trim();
23.         //使用 SharedPreferences 查询数据
24.         String result = sharedPreferences.getString(query_text, null);
25.         if(result == null) {
26.             tv_query.setText("您查询的数据不存在");
27.         }else {
28.             tv_query.setText(result);
29.         }
30.  }
31.  public void delete(View v) {
32.         //获得删除的键
33.         String delete = et_delete.getText().toString().trim();
34.         //使用 editor 删除数据
35.         editor.remove(delete);
36.         //一定要提交，该步骤非常容易被忽略
```

```
37.              editor.commit();
38.          }
39.      public void clear(View v) {
40.          //使用 editor 清空数据
41.              editor.clear();
42.          //一定要提交,该步骤非常容易被忽略
43.              editor.commit();
44.          }
45.  }
```

9.2.2 ProgressDialog

1. ProgressDialog 概述

在程序的执行过程中,有些操作可能需要较长时间,如某些资源的加载、文件的下载、大量数据的处理等,这时可以使用进度条告知用户明确的操作结束时间,让用户能够了解程序当前的进度及状态。利用 ProgressDialog(进度条对话框)可以实现上述目的,其主要用于显示操作的进度。

2. ProgressDialog 的常用方法

ProgressDialog 的使用比较简单,只要将其显示到前台,然后启动一个后台线程定时更改表示进度的数值即可。ProgressDialog 的常用方法及其作用如表 9-1 所示。

表 9-1　ProgressDialog 的常用方法及其作用

方　　　法	作　　　用
setProgressStyle()	设置进度条风格,如圆形
setTitlte()	设置ProgressDialog标题
setMessage()	设置ProgressDialog提示信息
setIcon()	设置ProgressDialog标题图标
setIndeterminate()	设置ProgressDialog的进度条是否不明确
setCancelable()	设置ProgressDialog是否可以按返回键取消
setButton()	设置ProgressDialog的一个Button(需要监听Button事件)
show()	显示ProgressDialog

9.2.3 Android 网络编程

1. HTTP 概述

HTTP 是一个属于应用层的、面向对象的协议,适用于分布式超媒体信息系统。
HTTP 的主要特点如下。
(1)简捷快速:当客户向服务器请求服务时,只需要传送请求方法和路径。
(2)请求方法常用的有 GET、HEAD、POST,这些方法规定了客户端与服务器联系的类

型，方法不同，联系的类型则不同。

（3）HTTP 协议简单，这使得 HTTP 服务器的程序规模小、通信速度快。

（4）灵活：HTTP 允许传输任意类型的数据对象。传输的类型由 Content-Type 加以标记。

（5）无连接：无连接的含义是限制每次连接只处理一个请求。

（6）服务器处理完客户端的请求，并收到客户端的应答后，即断开连接。

（7）采用 HTTP 方式可以节省传输时间。

（8）无状态：HTTP 协议是无状态协议，无状态是指协议对于事务处理没有记忆能力。缺少状态意味着如果后续处理需要前面的信息，则它必须重新传送，这样可能导致每次连接传送的数据量增大。另一方面，在服务器不需要先前信息时它的应答速度就较快。

2. Android 网络编程简介

下面介绍常用的 Android 网络开发技术。

（1）HttpClient。Android SDK 包含 HttpClient，6.0 版本的 Android 中则直接删除了 HttpClient 类库，如果仍想使用该类库，则可以通过以下方法实现。

如果使用的是 Eclipse，则在 libs 中加入 org.apache.http.legacy.jar 包，该 jar 包在 sdk\platforms\android-23\optional 目录中（需要下载 6.0 版本的 Android 的 SDK）。

如果使用的是 Android Studio，则在相应的项目的 build.gradle 中加入：

```
android {
    useLibrary 'org.apache.http.legacy'
}
```

（2）HttpURLConnection。对于 2.2 及之前版本的 Android，HttpURLConnection 存在一些 Bug，在此阶段使用 HttpClient 是较好的选择。而对于 2.3 及之后版本的 Android，HttpURLConnection 则是最佳的选择，其 API 简单，体积较小，因而非常适用于 Android 项目。另外，HttpURLConnection 的压缩和缓存机制可以有效减少网络访问的流量，在提升速度和省电方面也非常有优势。

（3）Volley。在 2013 年 Google I/O 大会上推出了一个新的网络通信框架，即 Volley。Volley 既可以访问网络并取得数据，也可以加载图片，并且其在性能方面进行了大幅度调整。Volley 非常适用于进行数据量不大但通信频繁的网络操作，而对于大数据量的网络操作，如下载文件等，Volley 的表现则非常糟糕。

（4）OkHttp。OkHttp 是目前应用较多的网络框架，它解决了很多网络问题，会从很多常用的连接问题中自动恢复。如果服务器配置了多个 IP 地址，当第一个 IP 地址连接失败时，OkHttp 会自动尝试连接下一个 IP 地址。此外，OkHttp 还解决了代理服务器问题和 SSL 握手失败问题。

3. HttpURLConnection

无论是开发者自己封装的网络请求类还是第三方的网络请求框架，都离不开 HttpURLConnection 类库。JDK 的 java.net 包提供了访问 HTTP 的基本功能的类：HttpURLConnection。

HttpURLConnection 是 Java 的标准类，继承自 URLConnection，可用于向指定网站发送 get 请求和 post 请求，它在 URLConnection 的基础上提供了如下便捷的方法。

int getResponseCode()：获取服务器的响应代码。

String getResponseMessage()：获取服务器的响应消息。
String getResponseMethod()：获取发送请求的方法。
void setRequestMethod(String method)：设置发送请求的方法。
HttpURLConnection 的使用步骤如下。

第一步：创建一个 URL 对象。

URL url = new URL(http://www.baidu.com);

第二步：利用 HttpURLConnection 对象从网络中获取网页数据。

HttpURLConnection conn = (HttpURLConnection) url.openConnection();

第三步：设置连接超时。

conn.setConnectTimeout(6*1000);

第四步：对响应代码进行判断。

if (conn.getResponseCode() != 200)   //从 Internet 获取网页，发送请求，将网页以流的形式读回来
    throw new RuntimeException("请求 url 失败");

第五步：处理输入/输出流。

InputStream is = conn.getInputStream();//获取输入流
String result = readData(is, "GBK"); //利用自定义方法 readDate()读取输入流中的数据
conn.disconnect();//数据连接

示例代码：

```
1.  public class NetUtils {
2.      public static String post(String url, String content) {
3.          HttpURLConnection conn = null;
4.          try {
5.              //创建一个 URL 对象
6.              URL mURL = new URL(url);
7.              //调用 URL 的 openConnection()方法，获取 HttpURLConnection 对象
8.              conn = (HttpURLConnection) mURL.openConnection();
9.              conn.setRequestMethod("POST");//设置请求方法为 POST
10.             conn.setReadTimeout(5000);//设置读取超时为 5s
11.             conn.setConnectTimeout(10000);//设置连接网络超时为 10s
12.             conn.setDoOutput(true);//设置此方法，允许向服务器输出内容
13.             //post 请求的参数
14.             String data = content;
15.             //获得一个输出流，向服务器写数据，默认情况下，系统不允许向服务器输出内容
16.             OutputStream out = conn.getOutputStream();//获得一个输出流，向服务器写数据
17.             out.write(data.getBytes());
18.             out.flush();
19.             out.close();
20.             int responseCode = conn.getResponseCode();//不必再使用 conn.connect()方法调用此方法
21.             if (responseCode == 200) {
22.                 InputStream is = conn.getInputStream();
23.                 String response = getStringFromInputStream(is);
24.                 return response;
25.             } else {
```

```
26.                throw new NetworkErrorException("response status is "+responseCode);
27.            }
28.        } catch (Exception e) {
29.            e.printStackTrace();
30.        } finally {
31.            if (conn != null) {
32.                conn.disconnect();//关闭连接
33.            }
34.        }
35.        return null;
36.    }
37.    public static String get(String url) {
38.        HttpURLConnection conn = null;
39.        try {
40.            //利用 String url 构建 URL 对象
41.            URL mURL = new URL(url);
42.            conn = (HttpURLConnection) mURL.openConnection();
43.            conn.setRequestMethod("GET");
44.            conn.setReadTimeout(5000);
45.            conn.setConnectTimeout(10000);
46.            int responseCode = conn.getResponseCode();
47.            if (responseCode == 200) {
48.                InputStream is = conn.getInputStream();
49.                String response = getStringFromInputStream(is);
50.                return response;
51.            } else {
52.                throw new NetworkErrorException("response status is "+responseCode);
53.            }
54.        } catch (Exception e) {
55.            e.printStackTrace();
56.        } finally {
57.            if (conn != null) {
58.                conn.disconnect();
59.            }
60.        }
61.        return null;
62.    }
63.    private static String getStringFromInputStream(InputStream is) throws IOException {
64.        ByteArrayOutputStream os = new ByteArrayOutputStream();
65.        //模板代码，必须熟练掌握
66.        byte[] buffer = new byte[1024];
67.        int len = -1;
68.        while ((len = is.read(buffer)) != -1) {
69.            os.write(buffer, 0, len);
70.        }
71.        is.close();
72.        //将流中的数据转换成字符串，采用的编码是 UTF-8（模拟器默认编码）
```

```
73.            String state = os.toString();
74.            os.close();
75.            return state;
76.        }
77. }
```

9.2.4　用 Android 原生技术解析 JSON

1. JSON 概述

JSON（JavaScript Object Notation）是一种轻量级的数据交换格式。因为解析 XML 比较复杂，而且需要编写大段代码，所以客户端和服务器的数据交换往往通过 JSON 进行。尤其对于 Web 开发来说，JSON 数据格式在客户端可以直接通过 JavaScript 进行解析。

JSON 有两种数据结构。一种是以 "key/value 对"（"键/值对"）形式存在的无序 JSONObject 对象，JSONObject 对象以{开始以}结束。每个 "键" 后跟一个 ":"；"键/值对" 之间使用 ","分隔。例如，{"name":"xiaoluo"}就是一个最简单的 JSON 对象，对于这种数据格式，key 必须是 String 类型，而 value 则可以是 String、Number、Object、Array 等数据类型。另一种数据格式是有序的 value 的集合，这种形式为 JSONArray，数组是值的有序集合。数组以［开始，以］结束，值之间使用 ","分隔。例如：

```
1. [
2.     {"id":1,"ide":"Eclipse","name":"java"},
3.     {"id":2,"ide":"XCode","name":"Swift"},
4.     {"id":3,"ide":"Visual Studio","name":"C##"}
5. ]
```

2. Android 的 JSON 解析部分主要用到的两个类

（1）JSONObject：可以看作 JSON 对象，它是系统中 JSON 定义的基本单元，其包含一系列 key/value 数值。在 JSONObject 对象中封装了 getXXX()等一系列方法，用于根据 JSON 对象中的 key 获取字符串、整型等类型的 value 值。JSONObject 类的 value 类型有 Boolean、JSONArray、Number、String 及默认值 JSONObject.NULLobject。

（2）JSONArray：代表一组有序的数值。将 JSONArray 转换为 String 输出（toString）所表现的形式是用方括号包裹，数值以 ","分隔（如［value1,value2,value3］）。JSONArray 封装的 get()方法可以通过 index 索引返回指定的数值，put()方法用来添加或替换数值。JSONArray 类的 value 类型有 Boolean、JSONArray、JSONObject、Number、String 及默认值 JSONObject.NULL object。

3. 用 Android 原生技术解析 JSON 的解读

例如，有一个 student 字段，其中包含了该 student 的一些基本属性，具体代码如下。

```
1. {
2.     "student":{
3.         "name":"rose",
4.         "age":"18",
5.         "isMan":true
6.     }
```

```
7.    }
```

在解析上面这段代码时,如果被{}包含,则为 JSONObject 对象,如果被[]包含,则为 JSONArray 对象。由此可以判断出上面这段代码的解析为 JSONObject 对象,其内部包含了一个 user 字段,该字段的值也是一个 JSONObject 对象。

```
1.  public class PullJSON {
2.      public static String json = "{\" student \":{\" name \":\" rose \",\"age\":\"18\",\"isMan\":true}}";
3.      public static void main(String[] args){
4.          JSONObject obj = new JSONObject(json);//最外层的 JSONObject 对象
5.          JSONObject user = obj.getJSONObject("student");//通过 student 获取其所对应的JSONObject 对象
6.          String name = user.getString("name");//通过 name 获取其所对应的字符串
7.          System.out.println(name);
8.      }
9.  }
```

上面这段代码的打印结果为 Rose。

9.3 热身任务

9.3.1 "我的进度条对话框"

1. 任务说明

"我的进度条"对话框效果图如图 9-2 所示。单击图 9-2(a)中的"圆形进度条"按钮,弹出如图 9-2(b)所示的圆形进度条对话框。单击图 9-2(a)图中的"长形进度条"按钮,则弹出如图 9-2(c)所示的长形进度条对话框。

(a)

(b)

(c)

图 9-2 "我的进度条对话框"效果图

2. 操作步骤

(1) 创建一个 Android 项目。

(2) 打开 activity_main.xml 文件，布局文件并完成布局效果，具体代码如下。

```
1.   <?xml version="1.0" encoding="utf-8"?>
2.   <LinearLayout xmlns:android="http://schemas.android.com/apk/res/android"
3.       xmlns:app="http://schemas.android.com/apk/res-auto"
4.       xmlns:tools="http://schemas.android.com/tools"
5.       android:layout_width="match_parent"
6.       android:layout_height="match_parent"
7.       android:orientation="vertical"
8.       tools:context=".MainActivity">
9.       <Button
10.          android:id="@+id/Button1"
11.          android:layout_width="wrap_content"
12.          android:layout_height="wrap_content"
13.          android:onClick="Dspinner"
14.          android:text="圆形进度条"/>
15.      <Button
16.          android:id="@+id/Button2"
17.          android:layout_width="wrap_content"
18.          android:layout_height="wrap_content"
19.          android:onClick="Dhorizaton"
20.          android:text="长形进度条"/>
21.  </LinearLayout>
```

(3) 打开 java 文件夹的 MainActivity.java 源程序，修改代码以实现相关功能，具体代码如下。

```
1.   public class MainActivity extends Activity {
2.       ProgressDialog xh_pDialog;
3.       int xh_count;
4.       protected void onCreate(Bundle savedInstanceState) {
5.           super.onCreate(savedInstanceState);
6.           setContentView(R.layout.activity_main);
7.       }
8.       public void Dspinner(View v) {
9.           //创建 ProgressDialog 对象
10.          xh_pDialog = new ProgressDialog(this);
11.          //设置进度条风格，风格为圆形、旋转的
12.          xh_pDialog.setProgressStyle(ProgressDialog.STYLE_SPINNER);
13.          //设置 ProgressDialog 标题
14.          xh_pDialog.setTitle("提示");
15.          //设置 ProgressDialog 提示信息
16.          xh_pDialog.setMessage("这是一个圆形进度条对话框");
17.          //设置 ProgressDialog 标题图标
18.          xh_pDialog.setIcon(android.R.drawable.ic_input_add);
19.          //设置 ProgressDialog 的进度条是否不明确，false 指不设置为不明确
20.          xh_pDialog.setIndeterminate(false);
```

```
21.        //设置 ProgressDialog 是否可以按退回键取消
22.        xh_pDialog.setCancelable(true);
23.        //设置 ProgressDialog 的一个 Button
24.        xh_pDialog.setButton("确定", new DialogInterface.OnClickListener() {
25.            @Override
26.            public void onClick(DialogInterface dialogInterface, int i) {
27.                dialogInterface.cancel();// 单击"确定"按钮取消显示对话框
28.            }
29.        });
30.        //显示 ProgressDialog
31.        xh_pDialog.show();
32.    }
33.    public void Dhorizaton(View v) {
34.        xh_count = 0;
35.        //创建 ProgressDialog 对象
36.        xh_pDialog = new ProgressDialog(this);
37.        //设置进度条风格,风格为圆形、旋转的
38.        xh_pDialog.setProgressStyle(ProgressDialog.STYLE_HORIZONTAL);
39.        //设置 ProgressDialog 标题
40.        xh_pDialog.setTitle("提示");
41.        //设置 ProgressDialog 提示信息
42.        xh_pDialog.setMessage("这是一个长形进度条对话框");
43.        //设置 ProgressDialog 标题图标
44.        xh_pDialog.setIcon(android.R.drawable.ic_input_add);
45.        //设置 ProgressDialog 的进度条是否不明确,false 就是不设置为不明确
46.        xh_pDialog.setIndeterminate(false);
47.        //设置 ProgressDialog 进度条进度
48.        xh_pDialog.setProgress(100);
49.        //设置 ProgressDialog 是否可以按退回键取消
50.        xh_pDialog.setCancelable(true);
51.        //让 ProgressDialog 显示
52.        xh_pDialog.show();
53.        new Thread() {
54.            @Override
55.            public void run() {
56.                try {
57.                    while (xh_count <= 100) {
58.                        //由线程来控制进度
59.                        xh_pDialog.setProgress(xh_count++);
60.                        Thread.sleep(100);
61.                    }
62.                    xh_pDialog.cancel();
63.                } catch (Exception e) {
64.                    xh_pDialog.cancel();
65.                }
66.            }
67.        }.start();
```

68.   }
69. }

9.3.2 "名人榜"

1. 任务说明

"名人榜"效果图如图 9-3 所示。当单击图 9-3（a）中的"中国名人榜"按钮后，可从自建服务器中的 celebrity.ashx 接口请求数据，若 celebrity.ashx 请求成功，返回的数据的 JSON 字符串（见图 9-4）接着通过 Android 原生技术解析出请求的 JSON 数据显示至 UI，如图 9-3（b）所示。

（a）

（b）

图 9-3　"名人榜"效果图

{'list':[{'name':'邓稼先','deed':'中国原子弹之父'},{'name':'华罗庚 ','deed':'中国现代数学之父'},{'name':'李四光','deed':'著名地质学家'},{'name':'周培源','deed':'著名物理学家'},{'name':'袁隆平','deed':'杂交水稻之父'},{'name':'钱三强','deed':'中国两弹之父'},{'name':'钱学森','deed':'中国航天导弹之父'},{'name':'苏步青','deed':'著名数学家'},{'name':'王淦昌','deed':'中国核武器之父'},{'name':'吴文俊','deed':' 著名数学家 '}]}

图 9.4　JSON 字符串

2. 操作步骤

（1）创建一个 Android 项目。

（2）打开 activity_main.xml 文件，布局文件并完成布局效果，具体代码如下。

```
1.  <?xml version="1.0" encoding="utf-8"?>
2.  <LinearLayout xmlns:android="http://schemas.android.com/apk/res/android"
3.      xmlns:app="http://schemas.android.com/apk/res-auto"
4.      xmlns:tools="http://schemas.android.com/tools"
5.      android:layout_width="match_parent"
6.      android:layout_height="match_parent"
7.      android:orientation="vertical"
8.      tools:context=".MainActivity">
9.      <ListView
10.         android:id="@+id/list"
11.         android:layout_width="match_parent"
12.         android:layout_height="wrap_content" />
13.     <Button
14.         android:id="@+id/button"
15.         android:layout_width="match_parent"
16.         android:layout_height="wrap_content"
17.         android:onClick="LoginThread"
18.         android:text="中国名人榜 " />
19. </LinearLayout>
```

（3）在项目\res\layout 文件夹中添加 list_item.xml 文件，用于规范每个选项元素的界面布局。list_item.xml 文件代码如下。

```
1.  <?xml version="1.0" encoding="utf-8"?>
2.  <LinearLayout xmlns:android="http://schemas.android.com/apk/res/android"
3.      android:layout_width="match_parent"
4.      android:layout_height="match_parent"
5.      android:gravity="center_vertical">
6.      <TextView
7.          android:id="@+id/name"
8.          android:layout_width="0dp"
9.          android:layout_height="wrap_content"
10.         android:layout_weight="2"
11.         android:textSize="20sp"
12.         android:textColor="#5CACEE"
13.         android:text="TextView" />
14.     <TextView
15.         android:id="@+id/deed"
16.         android:layout_width="0dp"
17.         android:layout_height="wrap_content"
18.         android:layout_weight="3"
19.         android:textColor="#66CD00"
20.         android:text="TextView" />
21. </LinearLayout>
```

（4）打开 java 文件夹的 MainActivity.java 源程序，修改代码以实现相关功能，具体代码如下。

```
1.  public class MainActivity extends Activity {
2.      ListView list;
3.      ArrayList arraylist;
```

```java
4.     private Handler handler = new Handler() {
5.         public void handleMessage(Message msg) {
6.             switch (msg.what) {
7.                 case 1:
8.                     String answer = msg.obj.toString().trim();
9.                     jsonParser(answer, "list");
10.                    SimpleAdapter adapter = new SimpleAdapter(getApplicationContext(), arraylist, R.layout.list_item, new String[]{"name", "deed"}, new int[]{R.id.name, R.id.deed});
11.                    list.setAdapter(adapter);
12.                    break;
13.                case 2:
14.                    Toast.makeText(getApplicationContext(), "网络异常，请重新连接", Toast.LENGTH_LONG).show();
15.                    break;
16.                case 3:
17.                    Toast.makeText(getApplicationContext(), "出现不明异常，请待会儿再试", Toast.LENGTH_LONG).show();
18.                    break;
19.                case 4:
20.                    Toast.makeText(getApplicationContext(), "json 解析不成功", Toast.LENGTH_LONG).show();
21.                    break;
22.            }
23.            super.handleMessage(msg);
24.        }
25.    };
26.    protected void onCreate(Bundle savedInstanceState) {
27.        super.onCreate(savedInstanceState);
28.        setContentView(R.layout.activity_main);
29.        list = this.findViewById(R.id.list);//查找 ListView 组件
30.    }
31.    //开启一个新线程，用于网络请求
32.    public void LoginThread(View v) {
33.        Thread login = new Thread(new Runnable() {
34.            public void run() {
35.                LoginHttpServlet();
36.            }
37.        });
38.        login.start();
39.    }
40.    public void LoginHttpServlet() {
41.        HttpURLConnection conn = null;
42.        InputStream is = null;
43.        try {
44.            // URL 地址
45.            String path = "http://172.20.51.99/celebrity.ashx";
46.            URL url = new URL(path);//得到访问地址的 URL
47.            conn = (HttpURLConnection) url.openConnection();//得到网络访问对象 java.net.HttpURLConnection
48.            conn.setConnectTimeout(3000); //设置超时时间
```

```
49.            conn.setRequestMethod("POST"); //设置获取信息的方式
50.            conn.setRequestProperty("Charset", "UTF-8"); //设置接收数据编码格式
51.            if (conn.getResponseCode() == 200) {
52.                is = conn.getInputStream();//获取数据流
53.                byte[] buffer = new byte[1024];
54.                int len = 0;
55.                StringBuilder sb = new StringBuilder();//创建 StringBuilder，用于拼接接收数据
56.                while ((len = is.read(buffer)) != -1) {
57.                    is.read(buffer);
58.                    String data = new String(buffer);
59.                    sb.append(data);
60.                }
61.                String response = sb.toString();
62.                Message msg = new Message();
63.                msg.what = 1;
64.                msg.obj = response;//将接收的数据捆绑至 message 中以便发送至 UI 线程
65.                handler.sendMessage(msg);
66.            }
67.        } catch (MalformedURLException e) {
68.            handler.sendEmptyMessage(2);
69.            e.printStackTrace();
70.        } catch (Exception e) {
71.            handler.sendEmptyMessage(3);
72.            e.printStackTrace();
73.        } finally {
74.            //意外退出时进行连接关闭保护
75.            if (conn != null) {
76.                conn.disconnect();
77.            }
78.            if (is != null) {
79.                try {
80.                    is.close();//关闭连接
81.                } catch (IOException e) {
82.                    e.printStackTrace();
83.                }
84.            }
85.        }
86.    }
87.    //JSON 解析数据，并存放至 ArrayList
88.    public void jsonParser(String jsonStr, String ArrayName) {
89.        /**这些是案例中请求到的数据
90.         {
91.         list:[{
92.         'name':'邓稼先','deed':'中国原子弹之父'
93.         },{
94.         name': '华罗庚 ', 'deed': '中国现代数学之父'
95.         },{
96.         'name': '李四光', 'deed': '著名地质学家'
```

```
97.            },{
98.              'name': '周培源', 'deed': '著名物理学家'
99.            },{
100.             'name': '袁隆平', 'deed': '杂交水稻之父 '
101.           },{
102.             'name': '钱三强', 'deed': '中国两弹之父'
103.           },{
104.             'name': '钱学森', 'deed': '中国航天导弹之父'
105.           },{
106.             'name': '苏步青', 'deed': '著名数学家'
107.           },{
108.             'name': '王淦昌', 'deed': '中国核武器之父'
109.           },{
110.             'name': '吴文俊', 'deed': '著名数学家 '
111.           }]}    **/
112.    try {
113.         arraylist = new ArrayList();
114.         //将请求到的 JSON 字符串转换成 JSONObject
115.         JSONObject jsonObject = new JSONObject(jsonStr);
116.         //从 JSONObject 抽取 list 这个 key 对应的 value，由于这个 value 是一个数组，因此
       把它转换成 JSONArray
117.         JSONArray jsonArray = jsonObject.getJSONArray("list");
118.         for (int i = 0; i < jsonArray.length(); i++) {
119.             //JSONArray 中的每个元素又是一个 Object，因此将数组中的每个元素转换在
       jsonArray 中的每个元素又是一个 JSONObject
120.             JSONObject obj = (JSONObject) jsonArray.get(i);
121.             String name = obj.getString("name");//提取 name
122.             String deed = obj.getString("deed");//提取 deed
123.             HashMap map = new HashMap();
124.             map.put("name", name);
125.             map.put("deed", deed);
126.             arraylist.add(map);
127.         }
128.    } catch (JSONException e) {
129.         handler.sendEmptyMessage(4);
130.         e.printStackTrace();
131.    }
132. }
133. }
```

第 88～132 行代码主要用于 JSON 字符串数据的解析，在进行解析时一定要清晰解析对象的结构。本案例对象可分解为如图 9-5 所示的形式。

 小贴士

由于 Android 中的网络操作也是一种耗时操作，因此在处理网络操作时，需要开启一个新的线程，然后利用 Handler 来实现 UI 线程及网络线程之间的通信。

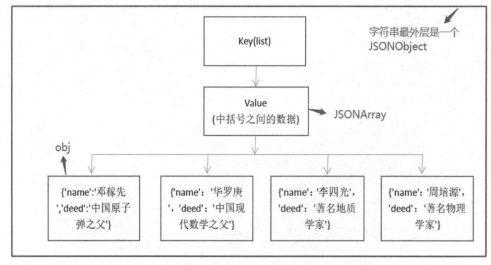

图 9-5 对象结构分解示意图

9.4 实现登录验证

1. 知识点

- Handler 通信机制。
- 利用 HttpURLConnection 进行 HTTP 网络通信。
- 利用 Android 原生技术解析 JSON 对象。
- ProgressDialog。

2. 工作任务

主要完成在"登录"界面（见图 9-6）输入用户名及密码，单击按钮后将用户名及密码上传至服务器，服务器对用户身份进行验证后，将验证结果返回客户端的功能。

图 9-6 登录界面

3. 操作流程

（1）打开项目中的 LoginActivity.java 程序，添加相应的功能代码，具体代码如下。

```
1.    public class Login extends Activity {
2.        EditText ETusername, ETpassword;//用户名及密码文本编辑框
3.        ImageView loginup;//"登入"按钮
4.        ProgressDialog dialog;
5.        private Handler handler = new Handler() {
6.            public void handleMessage(Message msg) {
7.                switch (msg.what) {
8.                    case 1:
9.                        dialog.dismiss();
10.                       String answer = msg.obj.toString().trim();
11.                       Toast.makeText(getApplicationContext(),answer, Toast.LENGTH_SHORT).show();
12.                       break;
13.                   case 2:
14.                       dialog.dismiss();
15.                       Toast.makeText(getApplicationContext(), "网络异常，请重新连接", Toast.LENGTH_LONG).show();
16.                       break;
17.                   case 3:
18.                       dialog.dismiss();
19.                       Toast.makeText(getApplicationContext(), "出现不明异常，让我理一理", Toast.LENGTH_LONG).show();
20.                       break;
21.                   default:
22.                       dialog.dismiss();
23.                       Toast.makeText(getApplicationContext(), "系统繁忙请稍后再试", Toast.LENGTH_LONG).show();
24.                       break;
25.               }
26.               super.handleMessage(msg);
27.           }
28.       };
29.
30.       protected void onCreate(Bundle savedInstanceState) {
31.           super.onCreate(savedInstanceState);
32.           setContentView(R.layout.activity_login);
33.           intView();
34.       }
35.       private void intView() {
36.           // TODO Auto-generated method stub
37.           ETusername = (EditText) this.findViewById(R.id.Login_editText1);
38.           ETpassword = (EditText) this.findViewById(R.id.Login_editText2);
39.           loginup = (ImageView) this.findViewById(R.id.Login_imageView1);
40.           loginup.setOnClickListener(new View.OnClickListener() {
41.               public void onClick(View arg0) {
42.                   String username = ETusername.getText().toString();
43.                   String password = ETpassword.getText().toString();
```

```
44.            if (username.length() != 0 && password.length() != 0) {
45.                //提示框
46.                dialog = new ProgressDialog(Login.this);
47.                dialog.setTitle("提示");
48.                dialog.setMessage("正在登陆，请稍后...");
49.                dialog.setCancelable(false);
50.                dialog.show();
51.                //创建子线程，分别进行 Get 传输
52.                LoginThread(username, password);
53.            } else {
54.                Toast.makeText(Login.this, "不能为空", Toast.LENGTH_SHORT).show();
55.            }
56.        }
57.    });
58. }
59. //开启新的线程进行 HTTP 网络通信
60. public void LoginThread(final String username, final String password) {
61.     Thread login = new Thread(new Runnable() {
62.         public void run() {
63.             LoginHttp(username, password);
64.         }
65.     });
66.     login.start();
67. }
68. //通过 HTTP 请求将用户名及密码上传至服务器进行验证，服务器将验证的结果返回，此方法将
    接收的数据通过 Handler 发送至主线程
69. public void LoginHttp(String username, String password) {
70.     HttpURLConnection conn = null;
71.     InputStream is = null;
72.     try {
73.         //用户名、密码
74.         // URL 地址
75.         String path = "http://172.20.51.99/login.ashx";//服务器验证程序地址
76.         path = path + "?username=" + username + "&password=" + password;
77.         conn = (HttpURLConnection) new URL(path).openConnection();
78.         conn.setConnectTimeout(3000); //设置超时时间
79.         conn.setDoInput(true);
80.         conn.setRequestMethod("POST"); //设置获取信息的方式
81.         conn.setRequestProperty("Charset", "UTF-8"); //设置接收数据编码格式
82.         if (conn.getResponseCode() == 200) {
83.             is = conn.getInputStream();
84.             byte[] buffer = new byte[1024];
85.             int len = 0;
86.             StringBuilder sb = new StringBuilder();//创建 StringBuilder，用于拼接接收数据
87.             while ((len = is.read(buffer)) != -1) {
88.                 is.read(buffer);
89.                 String data = new String(buffer);
90.                 sb.append(data);
91.             }
```

```
92.            String response = sb.toString();
93.            Message msg = new Message();
94.            msg.what = 1;
95.            msg.obj = response;
96.            handler.sendMessage(msg);
97.         }
98.      } catch (MalformedURLException e) {
99.         handler.sendEmptyMessage(2);
100.        e.printStackTrace();
101.     } catch (Exception e) {
102.        handler.sendEmptyMessage(3);
103.        e.printStackTrace();
104.     } finally {
105.        //意外退出时进行连接关闭保护
106.        if (conn != null) {
107.           conn.disconnect();
108.        }
109.        if (is != null) {
110.           try {
111.              is.close();
112.           } catch (IOException e) {
113.              e.printStackTrace();
114.           }
115.        }
116.     }
117.  }
```

 思考

使用 HTTP 网络请求还需要在 AndroidManifest.xml 文件进行什么设置才能实现其网络功能？

 9.5 实现登录信息本地保存

1. 知识点

➢ 使用 Android 原生技术解析 JSON 对象。
➢ SharedPreferences 储存方式的使用。

2. 工作任务

对服务器返回信息进行 JSON 解析，判断用户个人信息是否通过验证。若验证通过，则将用户名保存至 SharedPreferences 中，关闭登录界面，将"个人中心"的"待君登录"修改为服务器发回的用户名（见图 9-7（a）），同时以后此处用户名来源于 SharedPreferences；若验证未通过，则显示"用户不存在或密码错误"的消息框（见图 9-7（b））。

(a)　　　　　　　　　　　(b)

图 9-7　登录信息本地保存效果图

3. 操作流程

（1）打开项目中的 LoginActivity.java 程序，新建 save()方法，该方法通过 SharedPreferences 将个人信息保存至本地，具体代码及相关功能说明如下。

```
public void save(String username) {
    sp = this.getSharedPreferences("user_info", Context.MODE_PRIVATE);//创建 SharedPreferences
    SharedPreferences.Editor editor = sp.edit(); //创建 Editor
    editor.putString("username", username);//将数据保存至 UserName
    editor.commit();//提交业务
}
```

（2）新建 jsonParser()方法，该方法利用 Android 原生技术对服务器返回的 JSON 字符串进行解析，具体代码及相关功能说明如下。

```
public String jsonParser(String jsonStr) {
    /**
    用户信息合法时，服务器返回值：{"msg":"ok","UserName":"我是神人"}
    用户信息不合法时，服务器返回值：{"msg":"nook"}
    **/
    String user = "";
    try {
        //将请求到的 JSON 字符串转换成 JSONObject
        JSONObject jsonObject = new JSONObject(jsonStr);
        isok = jsonObject.getString("msg");//获取返回 JSON 对象中的 msg 的 value 值
        if (isok.equals("ok")) {
            //如果 msg 的值是 ok，说明验证通过，则获取 JSON 对象中的用户名信息
            user = jsonObject.getString("UserName");
        }
    } catch (JSONException e) {
```

```
            e.printStackTrace();
        }
        return user;
    }
```

（3）对程序中 Handler 的 case 1 分支进行编码，实现对 JSON 解析方法的调用及对服务器返回信息的判断，以确定用户是否登录成功，若登录成功则调用 save()方法进行储存，否则显示"用户不存在或密码错误"的消息框。Handler 修改后的代码如图 9-8 所示。

```
private Handler handler = handleMessage(msg) → {
    switch (msg.what) {
        case 1:
            String answer = msg.obj.toString().trim();
            String user=jsonParser(answer); //调用jsonParser方法将服务验证发回信息进行解析
            dialog.dismiss();
            if (isok.equals("ok") && user!=null) {
                save(user);
                Toast.makeText(getApplicationContext(), text:"登录成功", Toast.LENGTH_SHORT).show();
                finish();
            } else {
                Toast.makeText(getApplicationContext(), text:"用户不存在或密码错误", Toast.LENGTH_LONG).show();
            }
            break;
```

图 9-8 Handler 修改后的代码示意图

（4）打开项目中的 PersonalCenter.java 程序，在该程序中新建 initUser()方法，用于从本地 SharedPreferences 中读取数据并显示在"个人中心"的用户界面，具体代码如下。

```
public void initUser(View v) {
    sp=this.getActivity().getSharedPreferences("user_info",
android.content.Context.MODE_PRIVATE);
    String name=sp.getString("username","notsave");
    TVusername=(TextView) v.findViewById(R.id.cmember_textView1);
    if(name!="notsave"){
        TVusername.setText(name);
    }
}
```

（5）重写 Cmember_fragment.java 中的 onResume()方法，实现调用 initUser()方法的功能，具体代码如下。

```
public void onResume() {
    initUser(view);
    super.onResume();
}
```

SPOC官方公众号

欢迎广大院校师生 **免费注册体验**

www.hxspoc.cn

华信SPOC在线学习平台

专注教学

- 教学课件 师生实时同步
- 数百门精品课 数万种教学资源
- 多种在线工具 轻松翻转课堂
- 支持PC、微信使用
- 一键引用，快捷开课 自主上传、个性建课
- 测试、讨论 投票、弹幕…… 互动手段多样
- 教学数据全记录 专业分析、便捷导出

登录 www.hxspoc.com 检索 SPOC 使用教程 获取更多

SPOC宣传片

教学服务QQ群： 231641234
教学服务电话：010-88254578/4481   教学服务邮箱：hxspoc@phei.com.cn

———— 电子工业出版社有限公司   华信教育研究所 ————